System Engineering for IMS Networks

System Engineering for IMS Networks

Arun Handa

AMSTERDAM • BOSTON • HEIDELBERG • LONDON
NEW YORK • OXFORD • PARIS • SAN DIEGO
SAN FRANCISCO • SINGAPORE • SYDNEY • TOKYO

Newnes is an imprint of Elsevier

Newnes is an imprint of Elsevier
30 Corporate Drive, Suite 400, Burlington, MA 01803, USA
Linacre House, Jordan Hill, Oxford OX2 8DP, UK

Library of Congress Cataloging-in-Publication Data
Application submitted

British Library Cataloguing-in-Publication Data
A catalogue record for this book is available from the British Library.

ISBN: 978-0-7506-8388-3

For information on all Newnes publications
visit our Web site at www.books.elsevier.com

09 10 11 10 9 8 7 6 5 4 3 2 1

Printed in the United States of America

Working together to grow
libraries in developing countries

www.elsevier.com | www.bookaid.org | www.sabre.org

ELSEVIER   BOOK AID
            International   Sabre Foundation

Contents

List of Figures

List of Tables

Preface

The transformation of the mobile-phone has been one of the most profound innovations that we have witnessed over the past few years. It has morphed from simple voice communication equipment to an indispensable tool with a gamut of services. The steady stream of features and services deliver an experience beyond the basic two-way communication it evolved from. Interestingly, most of the experience we enjoy with the mobile-phone is only delivered by the device. A complex network, which works hard behind the scenes, is necessary to make this magic become real.

Over the years, the cellular network has evolved to provide the infrastructure enabling voice and data communications in a reliable and highly available service environment. The evolution of network technology has resulted in providing better voice quality, faster bandwidth, and even better handheld features. However, on the horizon is a technology that has the potential not just to provide an enhanced feature set, but to radically change the way we communicate. This is what makes the IP Multimedia Subsystem (IMS) exciting.

Telecommunication networks have become increasingly complex with the manifestation of innovative solutions from several disruptive technologies. What differentiates IMS is truly its ability to deliver as a *single platform* for what requires several methods today. Beyond that, IMS can be seen to unfold itself to deliver three promises. *Internet Protocol (IP)* brings forth the ability to connect any device to the wireless network. This could be a refrigerator, vending machine, or an alarm system. The possibilities are limitless. *Multimedia* translates to a new user experience. Most of us are familiar with voice, data, and video as distinct communication forms. IMS holds the potential to provide them in a blended paradigm. The *Subsystem* aspect reflects the ability of IMS to function as a single platform to achieve what is being offered with separate technologies. It also reflects the adaptive nature of IMS to co-exist with existing network infrastructures today and form a blueprint for converged networks.

IMS, as originally defined by the wireless standards, has gained significant industry momentum. Its acceptance has gone beyond the wireless networks that it was originally intended for, to cover wireline, broadband, and cable networks as well. This paradigm of convergence is being touted to change the telecommunication network within the next few years. Market Research from Analysts such as InStat, VDC, and ABI Research project this to be a multibillion dollar market in the next few years, providing service to several million subscribers.

IMS is defined as a reference architecture, which is specified across several standards documents. While the standards documentation is the last word in any implementation, these can be overwhelming, especially to the uninitiated. The standards are also evolving, which adds a level of complexity to track them. It is also hard to appreciate the ingenuity and innovation that has gone into conceiving the IMS by reading the specifications only. This book will help you overcome these challenges and gain a better understanding of the concepts and principles of IMS. However, I strongly recommend referring to the standards while building products and applications for IMS.

This book is intended to reach out to an audience whose goal is to engineer products for the IMS architecture. My aim is to provide a near self-contained text that builds the fundamentals for the unacquainted and carries across to product realization.

We will explore IMS from several angles. I will guide you through some basic building blocks, which will help you to understand the principles of a telecom network and explain how to realize applications in this new generation of networks.

The book is organized into four logical parts. In the first part we take an overview of IMS and build the necessary fundamentals about wireless and IP networks including protocols, which are required for an understanding of IMS. The astute reader may opt to skip these. We then move on to understand how it works. In the second part, we explore the various services and applications that are driving the business case for IMS. Next we examine the convergence of the wireless IMS with other networks. We then conclude how to engineer various products in the IMS network, which span from the handset to the core servers.

Consequently, this text will be useful to practitioners in these segments and also to engineering graduates looking for opportunities in this area, as this is not available as a course in most universities.

An Overview of What is Ahead

The introductory chapter gives the audience an overview of the IMS. The IMS was initially defined as a distinct phase in the wireless-3G evolution. Its feature of access-independence,

IP transport, and harmonization across other standards bodies has given it a significant acceptance from the telecom service providers. This chapter focuses on giving an overview of this reference architecture and what are the features that are molding this to be the base architecture for telecom networks for the next generation. We also discuss aspects of convergence and the overall vision of the Third Generation Partnership Project (3GPP), 3GPP2, European Telecommunication Standards Institute (ETSI)-Telecoms & Internet Converged Services & Protocols for Advanced Networks (TISPAN) and CableLabs standards that are driving IMS.

We start with a foundation of the basics of wireless and IP networks and protocols. IMS combines best-of-breed principles from both traditional wireless and IP networks. To the uninitiated, it is a little difficult to understand. I have observed that engineers in the traditional networks do not have a grasp of IP-related concepts and vice versa. These chapters help to explain how the telecom network architecture functions today describing the aspects of subscriber identity, registration, mobility, roaming, authentication, services, and so forth. for both Call-related and Data services. It then extends to the IP side to explain routing, addressing, Network Address Translation (NAT), IPv4, and IPv6, as these are fundamental to setting up IMS networks. Finally, we go through an overview of all the protocols that form the underpinning of the IMS network: Session Initiation Protocol (SIP), Diameter, MEGACO, and so forth.

My intent in the part is to open up the IMS black box and show the main functions inside and how they work. We start with the public and private identities of the users. Expand to the concepts of the Universal Subscriber Identity Module (USIM) and IMS Subscriber Identity Module (ISIM) and also understand about domains and realms in the IMS. We then extend to learning about the functions of Authorization, Authentication, and Accounting (AAA) giving an overview of the HSS and the subscriber profile. We then examine the control elements in the signaling space, which are namely the CSCFs, and understand their functions.

One of the important functions of IMS that differentiates it from the existing VoIP networks has been to guarantee the Quality of Service (QoS). The policy and methods that govern QoS are examined in this chapter. Service providers have been accustomed to inherent security from a closed network in their current deployed models. IP networks on the other hand have been seen to be more susceptible to hacker attacks. The IMS standards provide a good foundation for access and network security, which is also getting further fortified with solutions proven in Voice over Internet Protocol (VoIP) networks.

We then take a look at application servers and examine the Java, OSA/parlay, and IM-SSF approaches to put these together. We also look beyond standards to explore how to fold in applications from other networks. Service orchestration and the Service Capability Interaction Manager (SCIM) is a challenging topic, still evolving both in the industry and standards. This enables the co-existence of multiple services and their interleaving. Moving away from the signaling plane, we now examine the media elements and understand how voice and video are handled across the Media Resource Function Controller (MRFC) and Media Resource Function Platform (MRFP) functions. We explore online and offline charging methods, and finally to put it together, we walk through some scenarios to show how sessions are set up, services enabled, communication setup, and so forth.

In the next part we discuss various revenue-generating services that are driving IMS for consumer, enterprise, blended, and emergency services. We examine some case studies and walk through the message flow scenarios to see how they work.

Following that, we explore the convergence taking place with the evolution of the other networks. We will examine how IMS networks interact with other networks, and focus on the sister standards in the wireline and cable networks. We also look at the contemporary service approaches of the Service-Oriented-Architecture (SOA) and Service Delivery Platform (SDP), and discuss convergence with Web 2.0. These methods are not yet fully supported by standards, and offer an insight into innovative industry approaches.

In our final part, we will understand the technology issues to build products in the IMS space. These include factors for performance, cost, hardware, and software. We will examine software technologies that include operating systems and environments, and also hardware such as Advanced Telecom Computing Architecture (ATCA) and multi-core processing.

We look at how to develop IMS client-side applications. The handset side has stringent limitations in terms of the application footprint. The use of SIP clients effectively with compression techniques becomes essential. Further, we discuss the user interfaces, since the IMS applications require a complex interaction of graphic capability on a small screen.

We examine the control plane elements—the CSCFs and HSS—and see how to build a session control and transaction-oriented software architecture. We examine the various options to build media processing VoiceXML, Digital Signal Processing (DSP), and so forth. We then conclude with the methods to build application servers.

Standards

The IMS standards are maintained by the Third Generation Partnership Project (3GPP). The book is guided by these standards. I have made an effort to describe the book in terms of the Release 7 definition of the 3GPP standards, which have matured at the time of writing. Release 7 also harmonizes the view of an IP multimedia core from the other standards of the 3GPP2, ETSI-TISPAN, and PacketCable. The book uses the 3GPP standards as a primary reference, and explains the convergence with the other standards.

Acknowledgements

I owe the book to Harry Helms, my former acquisition editor at Elsevier. It was his confidence and faith in me that helped me to embark on this project. I wish him the best of health. I would like to thank Rachel Roumelioutis, my editor at Elsevier, for her unending patience and encouragement. My deep gratitude goes out to Alan Quayle, Chad Hart, Christophe Gourrad, and Gerry Christensen. They lent their IMS expertise to review and provided immense feedback to make the book where it stands. I would also like to thank my current and past colleagues at IntelliNet Technologies for their insight and comments and the opportunities to work in the IMS areas.

I dedicate this book to the memory of my mother, who I lost while writing this book; my father, who provided constant support. And finally my supporting wife Rachna and children Talin and Iksha, who bore with my weekend disappearances and missed out on a lot of family time.

I did learn that writing a book can be a back breaking exercise, literally speaking. It was, however, exciting for me to capture the thoughts in the industry and the knowledge acquired from the standards and practice. I hope you will find the book useful.

www.newnespress.com

CHAPTER 1

Introduction to IMS

The current generation of wireless cellular networks has relied on a digital network infrastructure to deliver traditional voice and data services. This provides the signaling and switching between the endpoints and various functional elements in the network. In the continued evolution path, wireless standards are focused on building the next-generation all-IP network infrastructure. This helps to provide higher bandwidths, lower the capital and operational costs of the network, offer a new generation of services, and provide seamless convergence with the Internet.

The IP Multimedia Subsystem (IMS) was crafted as a part of the third-generation (3G) network to solve these needs. The value of IMS has, however, grown beyond that. Both service providers and equipment vendors see the potential as a platform to solve other problems in the network. In this process of value-creation, IMS has different interpretations. Is it a technology, a platform, or a service-delivery framework?

1.1 What Is IMS?

According to the standards, IMS is defined in the form of a reference architecture to enable delivery of next-generation communication services of voice, data, video, wireless, and mobility over an Internet Protocol (IP) network. It is considered a subsystem, because it exists as part of a complete network. In other words, IMS by itself requires other components such as an access network, to fully function as a system for multimedia service delivery. It is considered a reference architecture, because the implementers build the functional elements conforming to these specifications. One interesting point is that as yet, a checklist has not been established to certify compliance or conformance.

Here are some key features of IMS that differentiate it from other network solutions.

> IMS is about providing a new user experience based on multimedia.
>
> IMS provides end-to-end SIP signaling.
>
> IMS is access network independent.
>
> IMS provides the mechanisms for controlling Quality of Service (QoS)
>
> IMS has the capability to deliver both bundled and combinational services.

Conceived from the vision of the 3G of cellular networks, the IMS has amassed its own independent status that now makes it the choice for implementing next-generation telecommunication networks. Initially, IMS was seen as a critical component in the evolution of the circuit-switched GSM (Global System for Mobile communications) network to an all-IP network. IMS is seen today as a versatile platform that can launch exciting applications that go beyond voice and data communications.

The standards on IMS are written and maintained by the third-generation partnership project (3GPP), which is a global partnership of organizations collaborating in 3G wireless technology standardization. IMS has, however, paved the way for other networks to evolve to an all-IP infrastructure. While purists may argue as to where IMS really belongs, the different networks in deployment today consider IMS as the next core network. Confluence with IMS is seen as the best option for converging with different networks, as shown in Figure 1.1.

In contrast to earlier standards that were implemented, the key differentiator in IMS that has enabled this harmonized concept across other standards is Access Network independence. The subsystem retains its own identity as long as an access can be provided by an IP connectivity network. The choice of transport becomes flexible. This has allowed the Code Division Multiple Access (CDMA), cable, and wireline standards to acknowledge the IMS as their IP multimedia platform.

3GPP first introduced the definition of the IMS architecture and its functional elements in their Release 5 for the UMTS (Universal Mobile Telecommunications System) Networks.

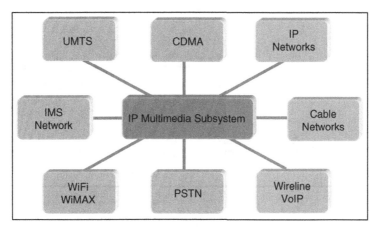

Figure 1.1 IMS as an enabler for convergence.

This was the first step toward the evolution of the circuit-switched core network to an all-IP core. 3GPP2, the standards body for the North American CDMA/ANSI-41 networks, adopted these standards as part of its Multimedia Domain (MMD) specifications.

As noted in Figure 1.2, the GSM radio networks with a circuit-switched core network have evolved to a UMTS/HSPA (High Speed Packet Access) radio network backed by a packet-switched core. This phased evolution as defined by the 3GPP introduced the IMS in Release 5 for the UMTS network standards. The initial release of IMS focused on defining the core functional elements. This laid the foundation of the IMS principles for session control, media processing, application services, access independence, and subscription, to name a few. Each release of 3GPP thereof continued to enrich these features and adapt them to the needs of the evolving network. The notable evolution aspect of IMS has been that the progress from Release 6 onward of the 3GPP standards continues to assimilate suggestions from all the peer standards. The most prominent ones have been the induction of the policy and flow-based charging and integrating Wireless Local Area Network (WLAN) and fixed broadband to provide voice call continuity. As service enablers and new services were introduced with IMS, it saw the induction of service definitions for Push-to-talk over Cellular, Presence, VideoSharing, and so forth. From a standards perspective, at the time of writing, Release 7 provides a stable base of specifications. Release 8 continues to define unaddressed areas such as Operations Administration and Maintenance (OAM) and service brokering, and focuses on optimizing the IMS for adapting it to the needs of fourth-generation (4G) networks.

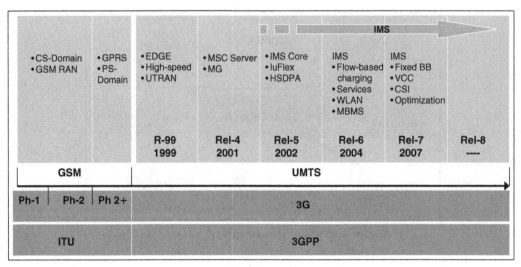

Figure 1.2 Standardization of IMS.

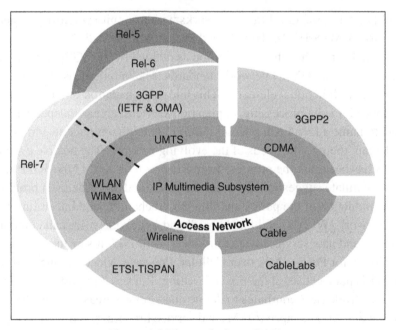

Figure 1.3 The evolution of IMS.

Figure 1.3 depicts the IMS as the core network for the multimedia domain for CDMA, Wireline, and broadband networks. The driving standards behind these networks—3GPP2, European Telecommunication Standards Institute (ETSI)-Telecoms & Internet Converged Services & Protocols for Advanced Networks (TISPAN), and CableLabs—have acknowledged IMS as their step to an all-IP evolution. The Open Mobile Alliance (OMA), which defines the service enablers for interoperable mobile devices, has also contributed to the specification of a rich service-enabling framework for the IMS network. IMS has been unique from its perspective of being able to harmonize and blend with multiple standards.

It is important to understand that IMS is not a technology—it is defined as a reference architecture. The principles that have been used for the definition of this architecture have been collated from best-of-breed solutions, as shown in Figure 1.4. These best practices, extended from both standards-based implementations and innovative approaches, have culminated in a system definition that inherits the perfect look.

Since IMS has been defined as an integral part of a cellular network, it inherits the concepts of mobility, roaming, and subscriber management from the proven execution models. The layered approach of separating the signaling (control-plane) and the traffic

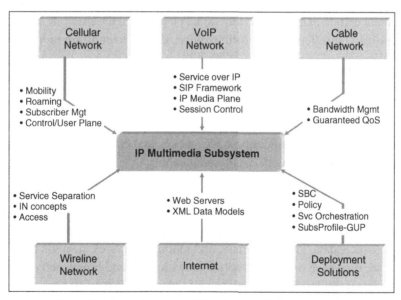

Figure 1.4 Inheriting principles from existing networks.

(user-plane) extends itself into the IMS network. While IMS uses these principles as a base, it acquires the concepts of IP-based communication from the Voice over Internet Protocol (VoIP) network. The capability to deliver a service over IP with a separate session control plane and a media plane using a SIP-based framework is modeled on the lines of the VoIP networks. It addresses the weakness of VoIP networks to provide guaranteed end-to-end QoS by service-based policy and better methods for resource reservation. It is influenced by policy management from cable networks, which help to deliver bandwidth management and QoS.

IMS has also derived the paradigm of the service plane separation and access network partitioning through Intelligent Network (IN). Further, IMS has applied effectively the concept of Web servers, eXtensible markup language (XML) data models, and presence services as they apply to the Internet.

But the most notable are the contributions that are still being fed into the definition of IMS, solving the challenges in deployments that are pronounced due to not being well addressed in the original standards. These are Session Border Control, Policy, Service Orchestration, and Subscriber Generic Profile to name a few.

A salient advantage of this approach is that existing telecommunication infrastructure can be leveraged to a certain extent. Some functional elements in the network such as user databases, session controllers, and so forth may be adapted to conform to this architecture.

What has really driven IMS in the limelight today is the ability of the telecommunication service providers to stay competitive in a disruptive, deflationary, and fragmenting landscape. Providing a telecommunication service is a capital-intensive business model. Unlike the venture capital-funded Internet, the telecom service providers have had to make their investments more judiciously. This means that change has to be better thought of and the evolution of the network clearly understood.

The core network, as we shall see in the next chapter, is evolving from a circuit-switched model to a packet-switched model. IP provides the signaling for this packet core network. Leveraging the core network in its current form, service providers are able to deliver voice, data, video, wireless, and mobility services in single or bundled forms. To do so, they have to resort to solutions with VoIP and broadband combinations. Triple-play systems can bundle voice, data, and video. Quad-play systems can add wireless capability. IMS is viewed as a single service-enabling platform that can launch

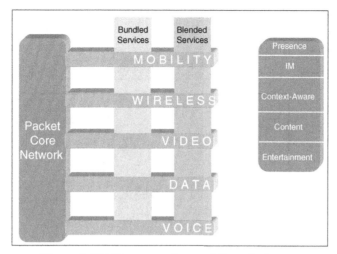

Figure 1.5 IMS as a service-enabling platform.

complete bundled services including mobility, and provide a new dimension of blended or combinational services. This enables a new generation of services with content, IM (Instant Messaging), presence, context-aware, and entertainment for voice-services dominated service providers, as shown in Figure 1.5. IMS potentially enables services across multiple access networks, providing adequate QoS controls are provided. So IMS potentially opens up new ways of structuring the business, which includes the transformation for both the wholesale and retail models.

1.2 A First Look at IMS

IMS is often presented with a complex diagram that is difficult to absorb at a first initiation. Let us take a two-step approach. First we examine IMS conceptually with the help of Figure 1.6, before we drill into details. As we discussed earlier, IMS is about providing the user with a new experience. IMS helps to deliver multimedia services, which include voice, data, video, IM, presence, and context on the handheld device. With its service architecture, IMS has the capability to provide these as disjoint services or blend them in a new combinational service. The communication between the handheld, which we will now refer to as the User Equipment (UE), and the IMS architecture, takes place over an access network. This is also referred to as the IP Connectivity Access Network (IP-CAN). As described earlier, the access network independence has been one

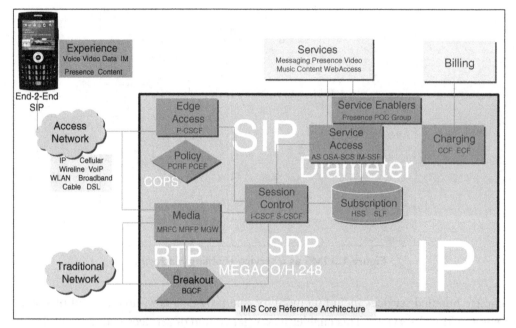

Figure 1.6 A first look at IMS.

of the major highlights of IMS, which has allowed convergence with various networks that exist today. As long as the access network can support IP signaling, it does not matter to IMS whether the access network uses wireless, wireline, broadband, or cable technology as the underlying transport.

All functional elements within the IMS architecture communicate with IP-based protocols. Session Initiation Protocol (SIP) is the most prominent, as it provides the capability to establish and control multimedia sessions. One reason why SIP is important, is that IMS is seen to provide a uniform protocol for signaling with the endpoints. This is in contrast to traditional wireless network implementation, where different protocols are used between the handset, access, and the core network. In other words, IMS provides end-to-end SIP signaling. It is, however, important to clear a misconception that IMS is not SIP, and SIP is not IMS. While SIP is the lingua franca of IMS, other protocols support vital functions as well. Diameter is the enabler for subscriber, policy, and charging functions. Megaco/H.248 and RTP provide the media-related support. COPS was used in the earlier IMS releases for policy functions, but has now given way to Diameter.

We now look at some of the functions provided by the various elements in the architecture. The Access function is an edge function, which is the entry point of signaling into the IMS network from the UE. This function, performed by the Proxy-Call Session Control Function, determines the home network to which the UE request be routed to. Policy is yet another important edge function, as it enables flow-based charging and bandwidth control for the access network. The control, coordination, and management of the multimedia session is performed by the Serving-Call Session Control Function (S-CSCF). The Interrogating Call Session Control Function enables getting to the right S-CSCF. The control of the session requires coordination with the subscriber functions, service access, and the media capabilities. Subscriber profiles enable the permissions and access for the sessions. Service Access functions provide the interface to the application servers, which direct the service logic for the sessions. Media Functions are controlled and delivered through the Media Resource Function Controller, Media Resource Function, and the Media Gateway. The Charging Function is implemented through the Charging Control Function and Event Charging Function. It gets event information from the elements in the architecture, and disseminates it to an external billing system.

To adapt the signaling to existing and traditional networks, the Breakout Gateway Control Function provides the inter-working to traditional networks.

1.3 The IMS Architecture

The IMS architecture is defined in terms of functional elements, their interaction, which is termed as reference points, and the protocols that carry out these interactions. The functional elements can also be logically grouped into layers or planes, based on the type of function they perform. This collection of the core functional elements has three distinct external interfaces to the user, application, and other networks. We will examine the IMS architecture from a Release 7 frame of reference for each of these attributes, as laid out in Figure 1.7.

1.3.1 Planes

The IP protocol provides the transport for the communication between the different functional elements. The information that flows in these IP packets is either of type signaling or media. The functional elements that participate in the signaling function fall into the control plane, as they provide the control function for establishing, maintaining,

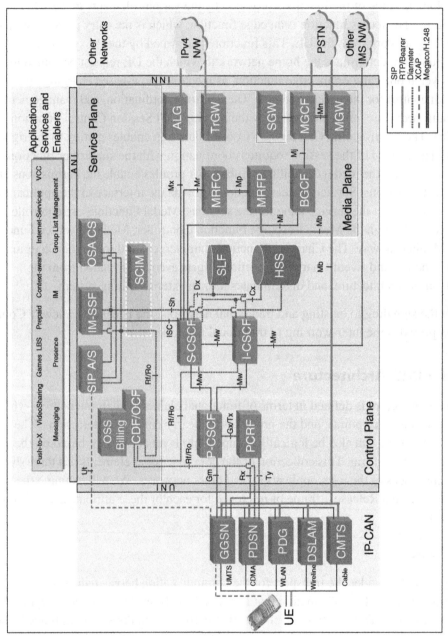

Figure 1.7 The IMS architecture.

and teardown of sessions. The media plane contains the elements that provide the resource and adaptation functions for physical media streams.

The service plane contains the functional elements, which host or provide an access to applications that enable service logic to the IMS user. These elements include application servers, service access functions, and Operational Support System (OSS) and Billing Servers.

1.3.2 Network Interfaces

There are three distinct interfaces between the entities the IMS network interacts with. These are the interfaces to the UE, to other networks, and to services and applications.

The User-Network Interface (UNI) is the interface to the IP-CAN, which is the access network between the user equipment and the IMS. Since IMS is agnostic of the access network technology, the interface points correspond to the gateway nodes in these networks. These elements are:

- Gateway GPRS Serving Node (GGSN) for UMTS networks

- Packet Data Serving Node (PDSN) for CDMA networks

- Packet Data Gateway (PDG) for Public Wireless LANs (802.11-based networks)

- DSL Access Modems (DSLAM) for VoIP and wireline broadband networks

- Cable Modem Termination System (CMTS) for the cable networks

This interface carries the SIP signaling toward the user equipment, diameter signaling to the wireless access entities for IP flow control, charging, and Real-Time Protocol (RTP) for the media streams.

The Network-Network Interface (NNI) corresponds to other networks, IP or non-IP, with which an IMS UE can communicate. This interface becomes complex, because the user terminals in the other network may not have an equivalent of a session or may rely on a non-IP technology. There are three types of networks to consider.

- A peer IMS network. This is a simple interface that relies on standard IMS signaling and media streams to a peer network. This is primarily to enable roaming between two IMS networks.

- An IP network that is based on the current IPv4 standard. The IMS standards define the use of IPv6, the next-generation IP standard for the transport network. Early IMS networks or SIP-based VoIP networks may use IPv4. Thus a conversion is required at the IP level to translate IPv4 to IPv4 signaling.

- The interface to the circuit-switched telecom network. The Public Services Telephone Network (PSTN) involves complexity. This requires protocol translation and inter-working on both the signaling and media planes. This requires a signaling gateway to translate between SIP and traditional telecom signaling such as IDSN User Part (ISUP), and a media gateway to adapt Time Division Multiplexing (TDM) voice to RTP streams.

The Application-Network Interface (ANI) has acquired a more complex view due to a partial definition by the standards, which is interpreted by implementers to adapt to their needs accordingly. The IMS standards specify a clear interface to SIP-based application servers and legacy services through gateway or inter-working functions. The capability of IMS to provide a greater value for combinational services has also made the case for the Service Capability Interaction Manager (SCIM) to play a stronger role as the interface to orchestrate between multiple applications and services. To summarize this interface involves the following models:

- Self-contained IMS applications that reside on SIP-based application servers.

- Applications hosted on SIP-based application servers, which interact with external servers, such as Internet server for exchange of information using Extensible Markup Language (XML), Simple Object Access Protocol (SOAP), or other standards.

- Interface to legacy services such as charging servers with the OSA parlay gateways.

- Interface to legacy services using the Customized Applications for Mobile Networks Enhanced Logic (CAMEL) or Intelligent Network (IN) protocols with the IMS Service Switching Function (IM-SSF) switching capability.

- Interface to the SCIM to provide an interaction between multiple services and applications.

1.3.3 IMS Functional Elements

1.3.3.1 Session Control

The Call Session Control Function (CSCF) provides the central control function in the IMS Core Network to set up, establish, modify, and tear down multimedia sessions (Figure 1.8). The CSCF function is distributed across three types of functional elements based on the specialized function they perform. These three elements are the Proxy CSCF (P-CSCF), Interrogating CSCF (I-CSCF), and the Serving CSCF (S-CSCF).

The P-CSCF is an edge access function and is the entry point for a UE to request services from an IMS network. The role of this CSCF is to function as a proxy by accepting incoming requests and forwarding them to the entity that can service them. The incoming requests are either the initial registration or an invitation for a multimedia session. A request for the UE to register for a service is normally forwarded to a session controller or to one with the capability to interrogate for it. Requests that are a session invitation are directed by the P-CSCF to a serving CSCF. The P-CSCF also performs some important edge functions. Since this is the first-hop access, it maintains a secure association with the UE. It also provides for compression of SIP signaling to minimize latency for over the air interface. It provides a policy function by initiating support for IP flow control and authorization of traffic-bearer resources. The P-CSCF is also capable of handling emergency call sessions.

The I-CSCF is responsible for determining which serving CSCF should be assigned for controlling the session requested by the UE. A request to the I-CSCF may come from the home network or a visited network through the proxy CSCF. The I-CSCF obtains the request for the address of the S-CSCF from the Home Subscriber Server (HSS) during a registration request, and provides it to the P-CSCF for subsequent multimedia requests.

The S-CSCF is responsible for conducting both registration and session control for the registered UE's sessions. It functions as a registrar and enables the network location information of the UE to be available at the HSS. It makes a determination to allow or deny service to the UE. It enables the assignment of application servers to the session, if required. Its role is to execute the session request by locating the destination endpoint and conducting the signaling toward it. The serving CSCF is also able to coordinate with the media resource function for any media announcements/tones to be played to the originating party. Serving CSCFs maintain a full state of the sessions and have the capability to originate and terminate a session on behalf of a requesting endpoint.

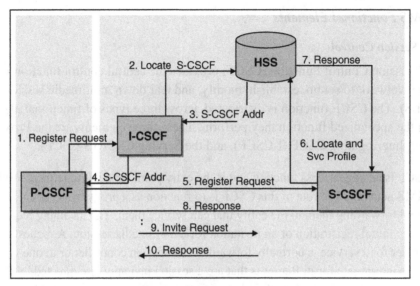

Figure 1.8 The Session Control.

All three CSCF functions are responsible for generating the Session Details or the Call Detail Records (CDRs). A typical network configuration may contain multiple instances of the CSCFs engineered to the appropriate traffic model.

The interface to the P-CSCF from the UE is the Gm interface that carries SIP/Session Description Protocol (SDP) signaling. The P-CSCF communicates to the interrogating and serving CSCF using SIP over the Mw interface.

1.3.3.2 Subscription

The Home Subscriber Server (HSS) can be viewed as a large database containing the complete subscription information about an IMS user. This information is accessible to all the IMS core elements needing information about the subscriber's profile, subscribed services, or authentication data. The HSS can be seen as a logical progression of the Home Location Register (HLR) in the second-generation (2G) networks. From an IMS-centric view, however, the HSS is required to support CSCFs and application servers. 3GPP takes a broader view of the HSS being able to support users of other networks as well. The HSS also has interfaces to legacy circuit-switched networks, packet-switched networks, and wireless LANs.

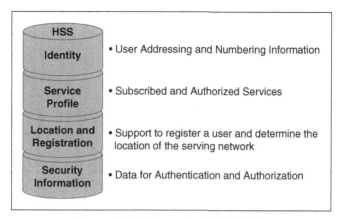

Figure 1.9 Information in the HSS.

The HSS stores and maintains the following information about a subscriber, as shown in Figure 1.9.

The HSS supports the CSCF functions by:

- Identifying the address of the CSCF that should be handling the session.

- Storing the user's registration and location information.

- Supporting the authentication and authorization by providing the integrity and ciphering data.

- Providing an access to a service profile, for which the subscriber has been provisioned.

The HSS also extends functionality to the application servers to determine service authorization, and also grants the capability to update subscriber profile data to application servers with provisioning capability.

The CSCF communicates with the HSS using Diameter over the Cx interface. The application servers use the Diameter Sh interface.

A large network may require provisioning the subscriber set into more than one HSS. This requires an intelligent entity to guide the requested CSCF or application server to the

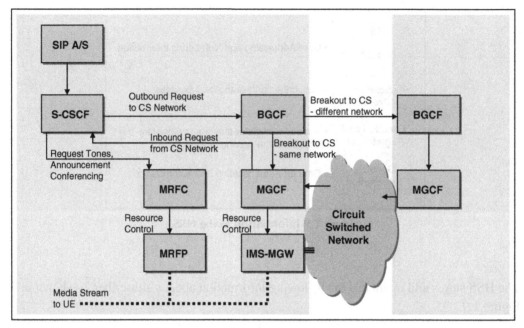

Figure 1.10 The media functions.

right HSS. The Subscriber Locator Function (SLF) provides this support to the I-CSCF during the initial registration and session setup. It extends this function to the S-CSCF during session control. It enables this for the application servers as well. The CSCFs request the HSS determination from the SLF over the Diameter Dx interface. The application servers use the Diameter Dh interface.

1.3.3.3 Media Functions

Having examined the elements in the control plane, we now examine the functional elements responsible for the multimedia in the media plane (Figure 1.10). The Multimedia Resource Function (MRF) encompasses the functionality to control the media stream and provide resources for processing it. The MRF comprises the Multimedia Resource Function Controller (MRFC) and the Multimedia Resource Function Processor (MRFP).

The bearer represents the actual multimedia stream carrying voice, data, and video. The MRFP provides the control of the bearer, which in the IMS core network is an

RTP stream. It provides the necessary resources for processing the media stream. The allocation and assignment of these resources is controlled by the MRFC. The processing of the media streams involves audio and video transcoding and analysis. To support multimedia conferencing, it provides the capability to mix multiple media streams and manage access to shared resources. It also supports the capability to play tones and announcements and can source a new media stream.

The function of the MRFC is to control the resource pool of the MRFP. The MRFC and MRFP have a master-slave relationship. The MRFC accepts the requests from the serving CSCF or an application server and controls the resources for the media stream accordingly.

The serving CSCF and the application servers can request media resources and services using SIP over the Mr interface. The MRFP controls the MRFC with an H.248 model over the Mp interface.

The MRF provides the capability to control and provide resources for multimedia streams over IP packets. The second set of elements in the media plane provides the capability to inter-work with the legacy circuit-switched (CS) network. Traffic is carried out on TDM streams. It is therefore essential to have a function to support media conversion between TDM and IP. This is performed by a Media Gateway (IMS-MGW). The control function has functionality split into a Breakout Gateway Control Function (BGCF) and a Media Gateway Control Function (MGCF) based on whether a call has to go outbound to the CS network or it is an inbound call from the CS network.

The IMS Media Gateway terminates the circuit-switched bearer channels and media streams from the packet network. It has to perform the necessary functions for media conversion and bearer control. Similar to the MRFP, it provides the necessary DSP resources for this function. In addition, it has to provide processing of the payload, which involves codecs, echo cancellation, and a bridge for conferencing.

The MGCF, similar to the MRFP, provides control of the MGW resources. In addition, it has to handle an inbound call from a CS network. It therefore needs to identify the right serving CSCF based on the routing number of the incoming call. The MGCF is also responsible for protocol conversion between the CS network ISUP signaling and SIP.

The serving CSCF does not make a request directly to the MGCF for an outbound call to the CS network. Instead, it requests the BCGF for determining the breakout from the

IMS network. What this means is to determine which CS network the call needs to be directed to. If the CS network is managed within the same operator domain, the BGCF will direct the request to the MGCF. If the BGCF identifies that the breakout has to occur to a different network, the BGCF will forward the signaling to the BGCF in that network.

The S-CSCF communicates with the BGCF using SIP over the Mi interface. The BCGF communicates with the MGCF using SIP over the Mj interface. The BGCF also communicates with a BGCF in another IMS network over the Mk interface. The MGCF controls the IMS-MGW with Megaco/H.248 over the Mp interface.

1.3.3.4 Service Functions

The service plane in the IMS is designed to support the next generation of application with SIP, and to be able to work with the Legacy service platforms. The service plane elements, referred to as the application servers, have the capability to support full service logic for an application. They can additionally function as a gateway or provide the inter-working function to a legacy server or a non-SIP server. Or the elements could coordinate service logic between multiple servers. Regardless of their role, all types of service elements communicate with the S-CSCF using SIP on the IMS Service Control (ISC) interface. They also have access to the subscriber information stored in the HSS.

The SIP Application Server (SIP AS) is a SIP server platform providing the value-added service logic to the IMS session control. The SIP AS can support services for call control, presence, and messaging to name a few. The SIP AS can also be used to inter-work with a non-SIP service, such as Web-based services.

Access to the OSA application servers is provided by the OSA Service Control Server (OSA SCS). The role of the OSA SCS is to inter-work between the ISC interface and the OSA interfaces. Similarly, access to the Legacy servers using CAMEL or the IN protocols require inter-working. Since the service logic in the IN-based networks is implemented as a service control function (SCF), the IMS Service Switching Function (IM-SSF) is defined to interact as an SSF function. Its primary role is to translate the ISC SIP requests to the IN protocols.

The Service Capability and Interaction Manager (SCIM) provides the support to create combinational service logic by the interaction between multiple application servers.

1.3.3.5 Charging Function

The IMS architecture provides a framework with several methods to monetize and charge for the services rendered through the subsystem. The functions support event-based, session-based, access-based, and flow-based charging methods. Charging can be applied with both online and offline charging methods. The architecture also utilizes the accounting mechanisms of the Authorization, Authentication, and Accounting (AAA) used for IP and data networks. This forms the basis for using Diameter in the various R, G, and W interfaces.

Offline charging provides the methods to record the service execution and resource usage information and store it for delivery to a billing function. This provides a straightforward mechanism to integrate with legacy and other third-party billing systems. The Charging Data Function (CDF) provides the necessary functionality to obtain the charging events and create CDRs for use by the billing system.

Online charging extends the capability to affect the service-execution in real time. This is based on obtaining the credit authorization for the access, volume, time, or other resource usage that is being requested for the service. The Online Charging System (OCS) provides the functionality for the decision logic, rating, and balance management to obtain the authorization for service execution.

The terms Charging Control Function (CCF) and Event Charging Function (ECF), which were defined by the initial IMS releases to support offline and online charging systems, respectively, are still in use.

Flow-based charging provides the mechanism to charge for the IP packets or IP flows, which are used in the session. This function is enabled at the IP-CAN layer. The service-based policy mechanism applied for QoS by the control and reservation of the IP flows provides a suitable mechanism to extend charging to the policy functions.

1.3.3.6 Policy

The capability of the policy functions to enable guaranteed end-to-end QoS and enable charging for it, likewise makes it a distinguishing feature of IMS. Traditional circuit-switched wireline/wireless networks did not face a QoS challenge. VoIP peer networks, on the other hand, strived for improving jitter and packet loss, but never set out an architectural principle to nail down a required QoS.

The service provider defines the policy to lay down the service rules and charging mechanisms. The execution of the policy function happens at the edge of the network. A UE requests for service and the edge devices need to determine what bandwidth can be allocated and whether any charging time-volume can be applied. Since the IMS network separates the control plane from the user or the traffic plane, application of policy rules requires coordination between the elements at the edge of the network.

The initial implementation of the policy function originated from the seeds of Service-based Local Policy (SBLP), with two functions, policy decision and policy enforcement. This has evolved to support charging correlation as well. The three entities that play a role in the policy function are:

- **P-CSCF** The Application Function (AF)

- **PCRF** The Policy Charging and Resource Function

- **PCEF** The Policy Charging Enforcement Function (the Gateway GPRS Service Node [GGSN]/Packet Data Serving Node [PDSN]).

1.3.3.7 Gateway Functions

The Signaling Gateway (SGW) is a border function that provides the signaling conversion between IP-based protocols and the legacy SS7 networks.

The Application Layer Gateway (ALG) and the Translation Gateway (TrGW) provide the support for the IP version inter-working between the IPv4 and IPv6 environments. The ALG provides application level translation at the SIP and SDP protocol level to communicate between IPv4 and IPv6 applications. The ALG can also be extended to support SIP/SDP for IPv4 VoIP-based networks. The TrGw performs the address and port translation.

1.3.4 Reference Points

Table 1.1 explains the various reference points or the interfaces between the functional elements.

Table 1.1 Reference points in the IMS architecture.

Interface	Between	Protocol	Function
Gm	UE ↔ P-CSCF	SIP	The Gm carries the SIP signaling between the UE and the P-CSCF. As part of it is over-the-air, it also requires signaling compression.
Mw	P-CSCF ↔ I-CSCF P-CSCF ↔ S-CSCF I-CSCF ↔ S-CSCF	SIP	The Mw signaling interface is used between the CSCFs for session setup, maintenance, and teardown.
Mr	S-CSCF ↔ MRFC	SIP	The Mr interface provides the support to the S-SCSF to request for media resources from the MRFC during the session control.
Mx	S-CSCF ↔ ALG	SIP	The Mx interface transports the signaling from the S-SCSF when the destination network requires IP version inter-working.
Mi	S-CSCF ↔ BGCF	SIP	The Mi interface transports the signaling from the S-SCSF toward a CS destination network that requires a breakout function.
Mj	BGCF ↔ MGCF	SIP	The Mj interface is used for the BGCF and MGCF interface within the same IMS core network.
Mk	BGCF ↔ BGCF	SIP	The Mk interface is used for the BGCF to forward the session to another BGCF in another IMS core network.
ISC	AS ↔ S-CSCF SCIM ↔ S-CSCF	SIP	The IMS Service Control interface supports the signaling for the transfer of control to an application server or the SCIM by the S-CSCF.
Cx	S-CSCF ↔ HSS I-CSCF ↔ HSS	Diameter	The Cx interface supports the functions for the (I/S) CSCFs to obtain and update user and service profile data from the HSS.

Continued

Table 1.1 (continued)

Interface	Between	Protocol	Function
Dx	S-CSCF ↔ SLF I-CSCF ↔ SLF	Diameter	The Dx interface is directed to the SLF from the (I/S)-CSCF to help locate the HSS in a multi-HSS network to serve a session request
Sh	AS ↔ HSS SCIM ↔ HSS	Diameter	The Sh interface supports the functions for the AS/SCIM to obtain and update user and service profile data from the HSS.
Dh	AS ↔ SLF SCIM ↔ SLF	Diameter	The Dh interface is directed to the SLF from the AS/SCIM to help locate the HSS in a multi-HSS network to serve a session request.
Rf	AS ↔ CDF MRFC ↔ CDF MGCF ↔ CDF BGCF ↔ CDF P-CSCF ↔ CDF I-CSCF ↔ CDF S-CSCF ↔ CDF	Diameter	The Rf interface provides the signaling support to carry the session and resource usage data for session-related events that can be provided to the CDF for creating offline accounting records.
Ro	AS ↔ OCS MRFC ↔ OCS MGCF ↔ OCS BGCF ↔ OCS P-CSCF ↔ OCS I-CSCF ↔ OCS IMS-GW ↔ OCS	Diameter	The Ro interface provides the signaling methods for control of service execution by obtaining credit authorization from the OCS and furnishing the necessary session and resource usage data.
Gx	P-CSCF ↔ PCRF	Diameter	The Gx interface is a 3GPP interface to enable the P-CSCF to request for resource reservation. The Tx interface is the corresponding 3GPP2 interface.

Continued

Table 1.1 (continued)

Interface	Between	Protocol	Function
Rx	P-CSCF ↔ GGSN	Diameter	The Rx interface is a 3GPP interface to apply policy and charging rules to the IP flows at the GGSN. The Ty interface is the corresponding 3GPP2 interface.
Mp	MRFC ↔ MRFP	Megaco/H.248	The Mp is an H.248 interface for controlling media resources on an MRFP by the MRFC.
Mn	MGCF ↔ IMS-MGW	Megaco/H.248	The Mn is an H.248 interface for controlling media resources on an IMS-MGW by the MGCF.
Mb	MRFP ↔ UE IMS-MGW ↔ UE	RTP	The Mb is the IP interface between the MRFP or the IMS-MGW and the UE carrying the IP packets corresponding to a media stream.
Ut	UE ↔ AS	XCAP	The Ut is an OMA-defined interface for the UE application clients to manage service-related data with the application servers.
Sr	MRFC ↔ AS	XCAP	The Sr interface is used by the MRFC to obtain information about the media such as scripts from the AS.

1.4 Chapter Summary

The IMS was originally conceived by the 3GPP wireless standards as its definition of an IP packet-based core network, which could deliver multimedia services. The evolution of the features of IMS has evolved from Release 5 to the current Release 7 and work-in-progress Release 8. The capabilities of this architecture have drawn the 3GPP2, TISPAN-NGN, CableLabs, and OMA to recognize this as the core architecture in their networks as well. IMS is thus viewed as the pivotal piece enabling convergence between multiple access networks and the Internet. IMS has also gained recognition as a versatile platform that can deliver a diverse set of services ranging from basic telephony to advanced multimedia services.

The most elegant feature of IMS—access network independence—is confusing to understand initially. IMS can deliver services to a user endpoint regardless of the IP connectivity network it chooses for access. IMS embodies the best-of-breed principles from various networks and consequently focuses on delivering an architecture with end-to-end SIP signaling with guaranteed QoS. This principle will enable a vision of a future network of a gamut of interconnected devices, not just mobile phones.

The IMS architecture is defined in terms of a set of functional elements and their interconnectivity identified by reference points. These elements can be logically grouped into three planes: session control, media, and the service. The session control and media plane are also referred to as the IMS Core Network as an abstraction.

CHAPTER 2

Basics of Wireless Networks

Looking at the IP Multimedia Subsystem (IMS) in isolation, it is difficult to find anything wireless about it. So why are wireless network concepts essential? The answer is that even though we see IMS as a blueprint for converged networks, IMS originated as and will continue to remain an integral part of the wireless network.

Two prominent standards define Wireless Telecom Networks. The International Telecommunication Union (ITU) and European Telecommunications Standards Institute (ETSI) standards have laid out the Global System for Mobile (GSM), General Packet Radio Service (GPRS), and Universal Mobile Telecommunication System (UMTS) networks. The American National Standards Institute (ANSI) standards have defined the ANSI41 and Code Division Multiple Access (CDMA), and Evolution-Data Optimized (EV-DO) networks. While we will focus more on the generic aspects of wireless networks, we will make some references to these two domains when required.

2.1 The Centralized Model of Wireless Telecom Networks

Before we delve deeper into wireless networks, let's examine the generic model of a wireless network in simple terms, as depicted in Figure 2.1. Wireless service providers or Mobile Network Operators (MNOs) aim to provide a service delivery model that comprises of the complete infrastructure a user requires to obtain a wireless service. In the US, the MNOs also provide the user device—the mobile phone or the data card (GPRS/EV-DO). Other MNOs globally give users the freedom to use their own device. In general, all MNOs prefer to own the rest of the infrastructure. This is implemented as a centralized model of control, where the intelligence is stored in the network. This is also referred to as a walled-garden, as it can be accessible in a limited form to the subscribers

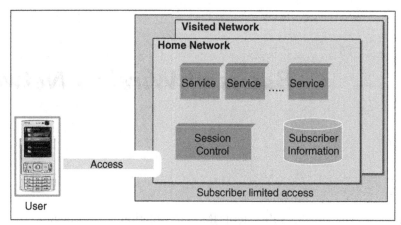

Figure 2.1 Generic wireless network model.

only. A simple security model is inherent in this approach. The protected and limited access provides a security perimeter, which makes it a trusted network.

The network is organized to provide the central element of the control of a user's interaction or a session. In traditional networks this has been termed a *call*, and in the next-generation networks is a *session*. All the subscription-related information is stored in a database, which is required for authorization, authentication, and service access based on the identity of the subscriber. Finally, there is the service logic, which defines how the services are to be delivered to the subscriber.

The Internet, on the other hand, has been built with an open access model. It delivers without centralized control by utilizing Peer-to-Peer (P2P) communication. Once an access is available to the network, the subscribers determine the control and behavior of their session. P2P models have been very effective for the Internet. Converging these two models has its set of challenges. These are to conform to government regulations, change in control model retained by the MNOs, and to implement stronger security mechanisms.

IMS leans toward the centralized model of control, with some flexibility to allow P2P interactions with supervision by the core and policy functions.

2.2 The RAN and the Core Network

A Telecom Network comprises two main parts: a Core Network (CN) and an Access Network (AN). The CN performs the functions for connection management, switching

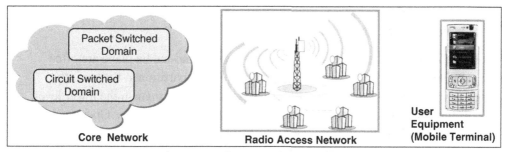

Figure 2.2 The CS and PS Domain.

user traffic and supporting signaling. It is characterized based on the type of switching it supports. Traditional networks have utilized circuit switching (CS) and are hence termed to be in the CS Domain. The support for Internet Protocol (IP)-based networks requires packet signaling; hence this is referred to as the PS Domain (Figure 2.2).

The AN, on the other hand, is the enabling network that allows a User Equipment (UE) or Terminal Equipment to get connectivity to the core network. In wireless networks, the access network is also referred to as the Radio Access Network (RAN). The RAN utilizes cellular or other technologies such as WiMax to provide the wireless access. The standardization of wireless networks has been done by ITU, ETSI, third-generation partnership project (3GPP) for GSM and UMTS standards and Telecommunications Industry Association (TIA) and 3GPP2 for CDMA standards. Each standard defines the infrastructure components in the CN and the RAN, and the protocols required for communication between these components.

2.3 Concept of the Control Plane and User Plane

Telecom networks have partitioned the end-to-end functionality between the UE and the CN into two distinct planes. The *control plane* is responsible for the signaling. This comprises the protocols required to effectively communicate between the components in the network. This is information required to set up, establish, and tear down a communication session between a UE and the network.

The *user plane* carries the actual media traffic. This encompasses voice, video, and data. Consequently, the user plane may also be referred to as the media plane or the data plane. Prior to the use of IP, the telecom networks used different communication channels for the control and the user plane. With IP, both signaling and media traffic packets are physically flowing through the same pipe.

The terms control plane and signaling plane; user plane and media plane are often used interchangeably. The network definition can also be partitioned further into two horizontal planes. The *transport plane* is also used to distinguish the lower level protocols that are primarily used for Layer 1 and 2 functionality of the Open Systems Interconnection (OSI) model, from the control signaling that takes place for Layer 3 and above. The *service plane* has been used to define the separation between sessions or call control with the logic that defines the delivery of a service to the user.

2.4 Functions of the Core Network

As we understood in the previous section, the core network elements perform the functions for connection management, switching user traffic and supporting signaling. In the CS domain, the main role of the core network has been to provide the following functions:

- Switching between the fixed networks and wireless radio networks. This is carried out by the Mobile Switching Center (MSC) or the Service Switching Function (SSF).

- Service control, to direct call behavior using service logic separated from the switching functions. This is done by the Service Control Point (SCP) or the Service Control Function (SCF).

- Subscription-related information, which provides authentication, service profiles, and user location. This is provisioned and maintained in the Home Location Register (HLR), Authentication Center (AuC), and Equipment Identity Register (EIR).

- Provisions for mobility and roaming between networks. This is done by the GMSC and Visitor Location Register (VLR).

- Messaging and other data services. These are the Short Messaging Service Center (SMSC) and GPRS/PDSN nodes.

IMS is a step in the evolution of the core network. While the transformation from circuit switching to IP does change the functional elements, the most core network principles remain preserved. The fundamental change is that the intensive function of switching call circuits gives way to multimedia session control. This is further simplified in IMS by using end-to-end SIP signaling. Service control in IMS is vested with application servers and the Service Capability Interaction Manager (SCIM). Subscription is handled by the Home Subscriber Server (HSS).

2.5 Subscriber Identity and Management

In a traditional wireless cellular network, UE such as a cell phone has two characteristics. One is that it has an identity as a device, and second, it holds the identity of the subscriber who owns and pays for the service of this device. The device identity, which is normally expressed in the manufacturer and equipment identifiers, helps in asset tracking or device capabilities. It is not as useful, however, as the subscriber identity. Wireless services are about the subscriber and not about devices. The subscriber identity is stored in the Subscriber Identity Module (SIM). This is hosted on a small chip that can be used interchangeably within different cellular devices in GSM/UMTS networks. In CDMA/ ANSI networks, this is integrated within the phone.

Subscriber identity in wireless networks has traditionally been used as a 10-digit number. This unique number is the public identity of the subscriber, which as we know is required to make or receive calls.

Each MNO provisions the subscribers in a database known as the Home Location Register (HLR). The HLR maintains responsibility for the user identification, numbering, and address. It enables network access control and authentication of this subscriber. It uses the identity to authorize or deny services, which are subscribed or not. It stores the network location of the user in the network the user is currently registered in.

2.6 Mobility and Roaming

Most users view mobility as the flexibility to move around unbound with a wireless phone. From a network perspective, it is really the ability to move freely between different networks in possibly different regions and with different MNOs. The beauty of the implementation of mobility in the cellular networks is that it is seamless and in most cases transparent to the user. With seamless handoffs between the access networks and the coordination with the core network infrastructure, mobility allows the user to roam freely. The mobility with cellular networks is a prime differentiator from other wireless technologies such as 802.11 and WiMAX, which have shown this in limited form.

To achieve mobility, we define a *home network*, where the user is provisioned. In this network, the HLR stores all the subscription-related information. The subscriber normally registers and obtains service within this home network. When the subscriber moves out of the home network and requests service of a different network, it is referred to as the *visited network*. The visited network has to have a roaming agreement with the home network, to be able to determine whether to provide service to this visitor. The following then needs to happen to enable roaming:

- The visited network must be able to determine the home network of the user.

- Consequently, it needs to obtain the authorization from the HLR.

- It then provides a temporary registration record to allow service.

- The HLR makes a note of the current location of the subscriber so that connection requests to the subscriber may be forwarded to the visited network.

2.7 Charging and Billing

Charging and billing are vital functions in the wireless telecom network that enable the mobile network to monetize the services provided. Charging is the process of obtaining the service and resource usage data and providing it to a billing system to apply tariff and pricing to bill the end user. Charging and billing systems are fairly complex. They are often considered a part of larger subscriber management systems referred to as Operational Support Systems (OSS) and Business Support Systems (BSS). Also viewed as a back-office operation, these systems need to work in conjunction in order to automate the charging/ billing process, service provisioning, and customer relationship management (Figure 2.3).

Different aspects of a wireless service can be monetized for charging purposes. A fee can be charged for network access, for instance. An event representing the unit of the type of service such as a call, data session, Short Messaging Service (SMS), or Multimedia Messaging Service (MMS), can be priced in terms of time or volume. Event- or session-based charging can be applied in an offline mode or online mode. In the offline mode, all the details related to the event such as time duration or resources used such as uplink/downlink bandwidth are saved as call detail records to be provided to a billing system. In the online

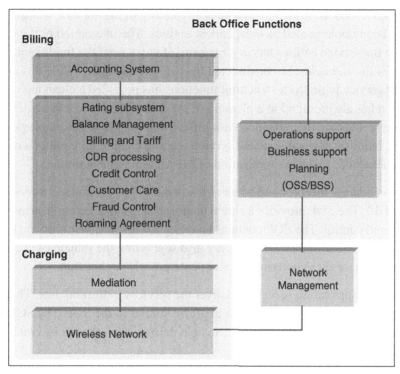

Figure 2.3 Billing and charging functions.

mode of charging, the call or data session is controlled through a rating and account balance function, which may affect the setup or the duration of the event. Offline charging lends itself well to postpaid billing, which is charging after the event has been completed. Online charging sets the base for prepaid charging. Online charging and prepaid work well together but are not the same. Online charging can also be used for postpaid billing.

Most billing systems are implemented as being agnostic to the underlying network technology. Charging in such systems requires a mediation system to translate between the network technology and the information required by the charging functions.

2.8 Service Delivery

Services delivered in the wireless network have traditionally been constructed as specialized elements that interact with the other core network functional elements.

Service elements such as SMSCs, voice mail, prepaid charging servers, or location centers have been implemented as independent entities. The interaction of the core network with the service entities has in a majority of cases used the Intelligent Network (IN) model for the call control to be directed by a service element. While this approach of separating service logic from switching functions has provided a clean logical architecture, it has also resulted in a plethora of service silos for the service provider. Moreover, these silos are implemented with different protocols and technologies. Furthermore, building new value-added services continues to require either a new specialized element or major customization of existing implementations.

One method to address this problem of service scalability has been to use Service Delivery Platforms (SDP). The SDP provides a single platform for service creation, provisioning, deployment, and control. The SDP contains an execution environment that can be common to a set of services. SDP implementations are also supporting the principles of Service Oriented Architecture (SOA) originating from enterprise IT architecture.

IMS relies on the application servers to deliver the services to the end user. The SCIM provides a strong base for a multi-service platform similar to the SDP. The SCIM with its capability to provide service interaction, can help deliver new services by combining the features from different services.

2.9 Network Management

The management of wireless telecommunication networks today uses a combination of the Telecommunication Management Network (TMN) and OSI standards. The OSI model is well accepted in managing most IP-based network elements. The TMN model provides the basic set of principles for a layered architecture and the management functional areas. The OSI model provides the management framework for the managing system and the management system.

TMN defines a layered architecture, which comprises four basic layers:

- **Element Management Layer** This layer deals with the management of the functions of a network element. Examples of these functions are detection of equipment errors, environmental conditions such as power consumption and temperature, resource measurement such as CPU, disk, and message queue utilization, and software upgrades.

- **Network Management Layer** This layer focuses on the interaction of the network elements and the topology of the network. Examples of the role of this layer are to configure the network topology, establish routing and signaling paths, and aggregating fault and measurement conditions across the network.

- **Service Management Layer** This layer is concerned with the aspects of the network that can be experienced by the users of the network. Examples of these network aspects are the Quality of Service (QoS)/Quality of Experience (QoE), accounting, user management, and so forth.

- **Business Management Layer** This layer focuses on the overall strategic management growth and evolution of the network.

The managing systems operate within the network management layer, and the managed systems are a part of the element management layer. The functional areas that are required to be supported by this framework are the following. These are also referred to as the FCAPS.

- Fault management

- Configuration management

- Accounting management

- Performance management

- Security management

There are no different standards for IMS network management. IMS network elements are expected to be able to support this standard methodology.

2.10 Chapter Summary

The essence of this chapter was to introduce some basic concepts in a wireless telecommunication network that are fundamental to IMS as well. The generic wireless network comprises the core network and a radio access network. The core network provides the functions of call/session control, subscriber management, authentication, and service delivery to the user device, and the RAN provides the wireless access. Mobility is an important concept, which allows a wireless phone to obtain service between different networks. The home network is where the mobile is provisioned, and the visited network is where the mobile requests service outside the home network.

For an MNO to monetize the technology in the core and the RAN, a set of essential support functions are required. The OSS/BSS systems provide the support for the operations support and business planning. These coordinate the billing and charging systems. Charging systems acquire the event information from the network and apply the necessary models of postpaid, prepaid, online, and offline methods.

CHAPTER 3

Basics of IP Networks

The choice of Internet Protocol (IP) for the backbone of the next generation networks has defined the DNA of IP Multimedia Subsystem (IMS). It signals a departure from closed networks and often proprietary, to open interconnected networks. IP has become the ubiquitous non-proprietary protocol suite, which is being used to communicate across any set of interconnected networks and is equally well suited for Local Area Network (LAN) and Wide Area Network (WAN) communications. While IP networks are engineered as WANs and LANs, access to the Internet as we use it today is administered by an access model. With this contrary approach to the telecom networks, a subscriber pays a fee to the Internet Service Provider (ISP) to access the network defined by some service characteristics such as bandwidth, downloads, and so forth. Mobile Network Operators (MNOs) see this approach as a mere bit-pipe approach, and view IMS as a counter to derive and deliver more service-related value.

While IP is a fundamental component of IMS, most low-level concepts are quite transparent. However, in this chapter we will cover the essentials that are required in engineering for IMS networks.

3.1 IP from an IMS Perspective

All telecommunications and internet network architecture, including IMS, closely follow the seven layer Open Systems Interconnection (OSI) model as shown in Figure 3.1. Layering helps to define a common set of functions that can be used for communication with peer entities. Often while designing network architecture, the lower protocol layers may be well abstracted if they have a mature implementation. Normally this is the case with IP protocols as well. The transport protocol set is quite transparent to an upper layer

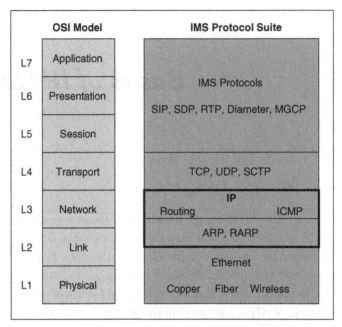

Figure 3.1 The IP suite in IMS.

such as the session or application. From an IMS perspective, we will not focus much on the transport layer; however, we do have to understand the principles of addressing and routing as they apply to the network layer or to the session/application layer.

3.2 Addressing

The current implementation of IP networks has been based on the IETF RFC 791 definition of the IPv4 protocol. Each entity in this network is identified by a unique 32-bit address. This is expressed as a four-octet (byte) scheme such as 192.168.1.100, which is illustrated in Figure 3.2.

IP addresses are further characterized into five classes of networks: Class A, B, C, D, and E. Mostly Class A, B, and C networks are used. Class D and E are reserved for multicast and experimental groups, respectively.

Figure 3.3 describes the allocation of IP addresses in the various classes of networks.

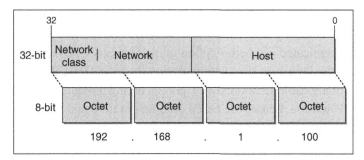

Figure 3.2 IP addressing.

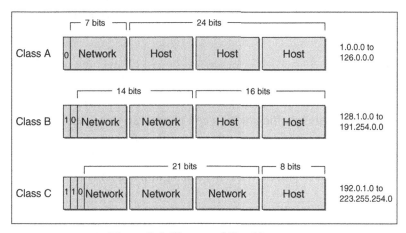

Figure 3.3 Classes of IP addresses.

Given this scheme, with a 32-bit address format, 2^{32} looks to be an impressive number to be able to handle 4 billion network devices. However, reflecting from Figure 3.3, administering a large network with 2^{24} devices, approximately 16 million does not seem like a very large number. Large MNOs will need to administer multiple such networks, especially when devices go beyond cell phones. This will not be easy. Later, we will see how IPv6 is suitable to mitigate this problem.

How are addresses allocated to endpoints? There are two methods: statically or dynamically. Static configuration of an address is to provision the address with an agreement with the network. Typically in the Internet, IP addresses are assigned and maintained by the InterNIC agency. In a private network, the corresponding Lightweight

Directory Access Protocol (LDAP) server or domain server is configured to recognize these addresses. Dynamically allocating these addresses when the device connects to a network and is authenticated is a better option to make judicious use of the address pool. This is done using the Dynamic Host Configuration Protocol (DHCP). A DHCP server maintains a pool of allocable IP addresses. Upon receiving a request from the device, the server leases an IP address. Leases are based on time duration and the device must renew the lease upon the aging of the lease.

DHCP is a recommended method in IMS for User Equipment (UE) to obtain access to the network. Since devices are mobile, they cannot be configured with a static IP address. The network providing service can use either DHCP to allocate the IP address or other General Packet Radio Service (GPRS)/Evolution-Data Optimized (EV-DO) methods, which we will study later.

3.3 IPv4/IPv6

One of the challenges we see is that the practical realization of addresses from a 32-bit addressing scheme appears insufficient for a cell phone-carrying population. Adding to it, other fixed IP endpoints would drastically reduce the efficacy of moving to an all-IP network. This was one of the problems IPv6 aimed to solve. Ratified nearly 20 years after IPv4, IPv6 proposed a few changes to the IPv4 datagram header (Figure 3.4).

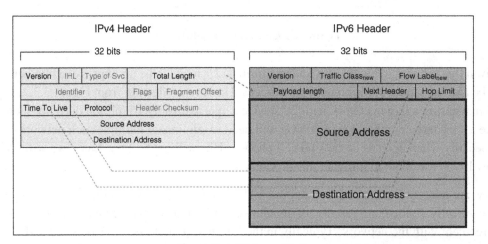

Figure 3.4 IPv4 vs. IPv6.

These changes are:

- Fourfold increase in address sizes: from 32 bits to 128 bits

- Twofold increase in header size: from 40 bytes to 80 bytes

- New fields added for Traffic Class and Flow Label to enhance Quality of Service (QoS)

- Some fields removed: IP Header Length, Type of Service, Flags, Fragment, Identification, and Header Checksum

- Hop limit replaced by Time To Live (TTL)

These changes brought the following improvements. With a fixed size header, removing the header length makes it easier for processing. The same holds true for byte alignment. The introduction of the next header field results in improved scalability and functionality. Counts based on hops are more effective for routing purposes rather than time-based metric (TTL). Header options to fortify for security are also provided.

IPv6 also introduces three types of addresses, which are quite suitable for next-generation communication paradigms suited for multimedia services:

- The Unicast, which identifies a single interface.

- The Anycast, which identifies a set of addresses in which the packet will be delivered to a single member of the set.

- The Multicast, which identifies a set of addresses to which the packet will be delivered.

Given these advantages, it is easy to see why IPv6 has been the recommended backbone for third-generation (3G) and IMS networks. However, practical realization is difficult. A vast legacy of IPv4 networks and equipment makes it challenging to start with a fresh IPv4 implementation. This results in the co-existence of IPv4 and IPv6 networks. There are two situations. In the first, an IPv4 network needs to communicate with an IPv6 network. This requires a translation device or a gateway at the edge of the network. The Stateless IP/Internet Control Message Protocol (ICMP) Translation Algorithm (SIIT) RFC-2765 provides a method to accomplish this. In the second situation, an IPv6 or an

IPv4 originating network has to route through an IPv4 or IPv6 network, respectively, to get to its homonymous destination. This problem can be resolved by tunneling, which can be accomplished either by the routers at the edge or created from the host to the end host.

3.4 Routing

Unlike circuit-switched networks, IP networks are designed to provide connectionless, best-effort delivery of the IP packets or datagrams. The determination of the path for an IP packet to reach its destination can be configured statically, but in internetworks, routing is established dynamically. Routes are calculated automatically by the routers in the path of the datagrams. This is made possible via IP routing tables, which define the path in terms of the destination address and the next hop pairs. When the datagrams initiate their travel, the entire route to get to the destination is not pre-determined. At each hop, the current node's routing table is examined for the next hop corresponding to the entry in the table matching the destination address of the datagram. The nodes are not responsible for the datagrams to reach their destination. This is coordinated by the ICMP.

In their transmission path, networks may have data links designed to support different maximum transmission unit (MTU) sizes. IP provides the capability to fragment, route, and reassemble datagrams.

3.5 Trusted Networks

As we observed in Figure 2.1, circuit-switched wireless networks have maintained an inherent security framework. Access to these networks is limited to private subscribers and network components are opaque. IP networks in contrast, primarily the Internet, have maintained public access and transparency. On the downside, this has led to the familiar problems of viruses, worms, and hack attacks, which render even private networks vulnerable. Since IMS uses an IP network, applying the necessary security rules and policies becomes essential to fortify the network.

A network security policy defines the rules for the usage of the network and the traffic that flows through it. It identifies a boundary or the perimeter of the network, where the policies can be enforced to protect the network resources and guard against the threats. The network that resides within the security parameter is referred to as a *trusted network*. There are three steps required to build a secure or a trusted network, as shown in Figure 3.5.

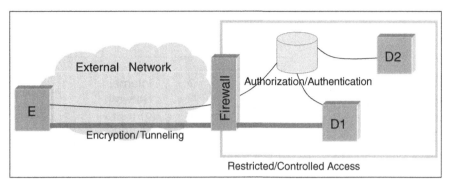

Figure 3.5 Trusted network.

1. Implement a contained network with controlled access to an external network. This is done by defining a Class-B or Class-C type network, which is confined in the bounds of a firewall. A firewall provides restricted access to the external network, which can be extended to different security levels.

2. Ensure authorized and authenticated access. This implies that the entities within the network or within an external peer network are known to have a trust relationship. This either requires obtaining authorization from a database, or authenticating them via the use of certificates or key exchanges.

3. Enforce secure communication between the entities. Encryption using Internet Protocol Security (IPSec) or tunneling using Virtual Private Network (VPN) methods allows the communication especially to a peer entity, while communicating over a public access network.

3.6 NAT/Firewall Traversal — Session Border Control

Network Address Translation (NAT) has been an effective technique used to insulate a private internal IP network and map these to a single external IP address. This function is normally implemented on an edge device, and is mostly combined with a firewall function. This method of enabling private networks while giving them access to the public internetworks helps conserve the pool of IP addresses, which are quite limited. NAT enables the address translation using an external port number for each connection. The 16-bit port address space extends the IP addressing capability to augment 65K potential assignments to a single IP address. We see this in Figure 3.6. Consider a typical

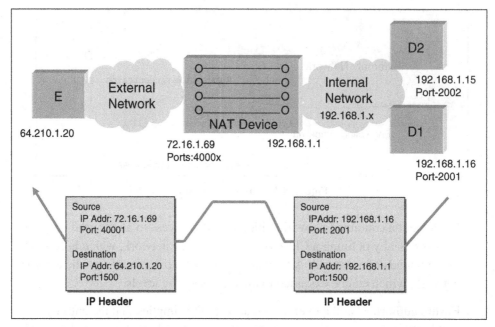

Figure 3.6 Network Address Translation.

internal Class-C network 192.168.1.x. D1 is an IP device that needs to make a session request to an external network entity E. Since E is not accessible to D1, the request goes to its internal firewall 192.168.1.1. The NAT function translates this request to the external entity. It maps its external IP address 72.16.1.69 and a port 40001 from its pool.

NAT seems easy and quite effective in solving our address limitation, so what is the problem? If we recollect from Figure 3.1, the focus of NAT has been toward the network stack layer. The application layers of the protocols pose a challenge. Most application protocols exchange address-related information about the source or the destination, especially to enable the return response. The problem is that an internal address or external address used will be unknown to the device on the other side of the firewall. The solution requires being able to open application layer packets and do the similar translation. This function is performed by an application level gateway (ALG).

The most well known ALG today is the session-border-controller (SBC). The SBC is an edge network device that was designed to solve the NAT traversal problem for SIP in

Voice over Internet Protocol (VoIP) networks. The SBC has to open the SIP packets and perform address translation.

It was believed that with IPv6, unlimited IP addresses will proliferate and the NAT traversal problem will go away and so will the need for SBCs. Given the advantages of secure private networks and acceptance of IPv4 interim for IMS, NAT traversal is still a challenge and SBCs still continue to be a viable solution.

3.7 Chapter Summary

We examined some essential concepts of IP networks that are required for engineering IMS networks. These focus on the addressing and routing. The current implementation of IP networks, referred to as IPv4, uses a 32-bit address space. This becomes a limitation while connecting a large set of devices. IPv6 overcomes this limitation and proposes a set of other improvements for improving quality of service and multicasting support. IPv6 is the recommended backbone for IMS. Its limited availability so far has pushed for working with IPv4 and resolving the necessary constraints. NAT traversal is one such problem, which disrupts an IP application signaling packet such as SIP, containing low-level address information to be interpreted correctly when traversing through an internal and external network.

CHAPTER 4

The IMS-related Protocols

The building blocks of the IP Multimedia Subsystem (IMS) network are a set of Internet Protocols (IP). Most of these have been proven and deployed in IP communication infrastructure. As we see in Figure 4.1, the protocols conform to the IP hourglass model corresponding to the seven-layer Open Systems Interconnection (OSI) network model. Our focus in this chapter will be mostly on the session and application layer protocols. We will also examine some adjunct protocols in the context of IMS.

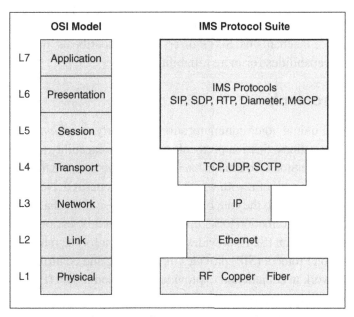

Figure 4.1 The IMS protocol suite.

4.1 Transport Protocols

Three transport protocols are used in IMS: Transmission Control Protocol (TCP), User Datagram Protocol (UDP), and Stream Control Transmission Protocol (SCTP). TCP provides reliable connection-oriented delivery of packets, and is suitable for the session-oriented protocols, which we will see in the next section. UDP provides unordered delivery of IP datagrams, which find an excellent application in carrying multimedia traffic. Due to the real-time nature of sampled audio and video data, the TCP features of retransmissions, flow control, and reordering are not appropriate. Finally, SCTP provides more reliability than TCP by better means of flow control, buffering, retransmission, and multiple streams. SCTP finds applicability in the IMS Authentication, Authorization, and Accounting (AAA) protocols and in the border function of the Signaling Gateway (SGW) interfacing to the Public Switched Telephone Network (PSTN).

The transport protocols provide services to their upper layers at well-defined interface points, which are also referred as *ports*. The IP address and the port are an important combination to set up a transport, connection, or stream.

All three protocols have the capability to multiplex on the ports. The transport protocols have an inherent capability for checksums, which allow detecting corrupted packets. TCP and SCTP extend additional reliability by retransmission of lost packets and by providing congestion avoidance mechanisms. SCTP offers built-in heartbeats, better flow control, and multi-homing capabilities for more reliability.

4.2 Session Protocols

A *call* is the familiar unit of voice communications. Similarly, a *session* is the unit of multimedia communications. Session protocols provide the capability to create, modify, extend, and terminate sessions. To establish a multimedia session, we first need to determine and locate the endpoint with which the session is desired. Next, the description of the session is to be delivered to the endpoint describing the multimedia capabilities. Finally, the exchange of control information to set up and tear down this session is necessary. The Session Description Protocol (SDP) provides the description for the multimedia capabilities. The Session Initiation Protocol (SIP) provides the signaling and control for the session, thus both SIP and SDP work in conjunction to provide this session model (Figure 4.2).

SIP messages are carried in TCP or UDP packets. SDP is carried as a payload in a SIP message such as an INVITE to start, modify, or extend a session.

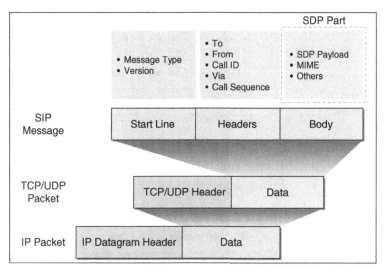

Figure 4.2 The SIP and SDP packets.

4.2.1 SDP

SIP uses SDP in an offer/answer paradigm. The originator of the session *offers* a description or a set of capabilities for the multimedia. The destination *answers* with a new description based on its capabilities. Once both sides have negotiated their view of the session, it can then be established. SDP provides two categories of description: the session-level information, which is relevant to SIP, and media capabilities. Session-level information includes version and connection-level information. Media capabilities include the type of media, such as audio/video, transports, and codecs.

SDP is an American Standard Code for Information Interchange (ASCII) text-based description, described in the form of lines (Figure 4.3). Each line in SDP provides a description for the session, time, and media-related information. Table 4.1 describes the options that are permissible.

Let's examine an SDP description to relate to the above. The following is an example in which a user named Brad from an IP endpoint 68.254.18.1 wants to establish a multimedia session for auction bidding, with both audio and video capability.

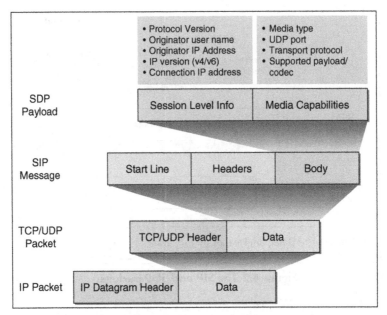

Figure 4.3 The Session Description Protocol.

Session information	v = 0	Protocol version
	o = Brad 2890844526 2890842807 IN IP4 68.254.18.1	Owner of the session and session ID
	s = Auction Bid	Name of the multimedia session
	c = IN IP4 68.254.18.1	Connection Information
	t = 0 0	Time the session goes active.
Media capability	m = audio RTP/AVP 0 97	Media. Audio in this case, 0 corresponding to G.711 μ-law
	a = sendrecv	Attribute for the media, which is bidirectional.
	m = video 49232 RTP/AVP 34 31	Media. Video in this case, 34 and 31 corresponding to video codecs H.263 and H.261

Table 4.1 SDP attributes.

Type	Meaning	Occurrence
v =	Protocol version	Required
o =	Owner and session ID	Required
s =	Session name	Required
i =	Session information	Zero or more
u =	URI of description	Zero or more
e =	E-mail address	Zero or more
p =	Phone number	Zero or more
c =	Connection information	Zero or more
b =	Bandwidth information	Zero or more
t =	Time description	Zero or more
r =	Repeat times for the session	Zero or more
z =	Time zone adjustments	Zero or more
k =	Encryption key for the session	Zero or more
a =	Session attributes	Zero or more
m =	Media name and transport address	Zero or more
i =	Media title	Zero or more
c =	Connection information	Zero or more
b =	Bandwidth for the media	Zero or more
k =	Encryption key for the media	Zero or more
a =	Attributes of the media	Zero or more

Session Description (v = through a =); Time Description (t = through z =); Media Description (m = through a =)

4.2.2 SIP

SIP is a session layer signaling protocol. It is also referred to as a Rendezvous protocol, because it is used to establish, modify, and terminate multimedia sessions. SIP is capable of using any of the IP transports—TCP, UDP, SCTP—and can work with Transport Layer

Security (TLS). SIP is a text-based protocol, which closely resembles Hypertext Transfer Protocol (HTTP). Similar to Uniform Resource Locators (URLs), it uses Uniform Resource Indicators (URI) to identify a SIP endpoint.

SIP was proposed for a standard in 1999 by the Internet Engineering Task Force (IETF) with RFC-2543. With substantial revisions, it was re-released as RFC-3261. In addition to this, there are several SIP extensions including significant ones for presence and messaging, which have been published as Request for Comments (RFCs). SIP implementations have matured since then. The adoption of SIP in Voice over Internet Protocol (VoIP) networks also provides a stable platform for use in IMS.

4.2.2.1 SIP Elements

In a SIP network, there are two kinds of SIP endpoints: a SIP User Agent (UA) or a SIP Server. A UA may have minimal functionality to be able to initiate and establish a session. It may have advanced features for conferencing, presence, and IM. A UA may also serve a specialized function such as a gateway to the PSTN.

SIP servers provide specialized functions to a UA, as shown in Figure 4.4. A User Agent Server (UAS) is a simple service invocation server. A Registrar server accepts a request from the UA to record its network location for routing sessions to it. A Proxy server receives a request from a UA and provides the capability to modify session data and extends the session on the behalf of the originating UA to another UA. It can perform a Domain Name System (DNS) lookup to determine where to route to or look up some service data to determine any modification to the session data. It can either be stateless or stateful. A UA request can be routed to a Redirect server, for a determination of where this request should be forwarded (for instance, to voicemail). The Back-to-Back User Agent (B2BUA) terminates a SIP session and then re-originates a new session, while maintaining an association between the two. Note the difference between the proxy and the B2BUA about creating a new session. Also, a proxy deals with signaling only. It ignores the SDP message body and does not see any user traffic flow. A B2BUA, on the other hand, has a full message body handling capability and in most cases sees user media traffic (RTP). B2BUAs find applications in applications such as prepaid controllers.

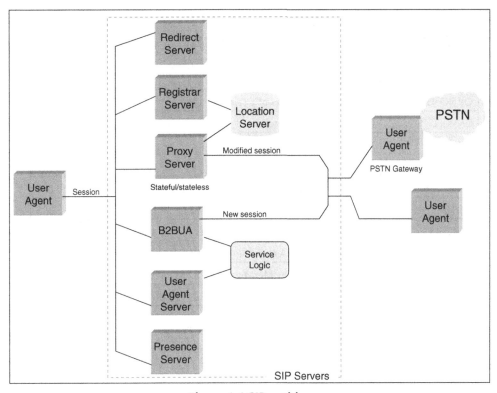

Figure 4.4 SIP entities.

4.2.2.2 SIP Addressing

SIP uses the familiar Internet application addressing, which is of the form *user@domain*. A SIP URI is of the form

sip:johndoe@somewhere.com	The regular SIP URI
sips:johndoe@somewhere.com	The secure SIP URI, which uses TLS over TCP
tel:3217955695@somewhere.com	For telephone numbers
pres:pserver@somewhere.com	For a presence resource
im:iserver@somewhere.com	For an instant messaging resource

Table 4.2 SIP commands.

Method	Request purpose	Reference
INVITE	Establish a session with offer/answer	RFC 3261
ACK	Acknowledge a response to an INVITE	RFC 3261
OPTIONS	Query the capabilities of a server or UA	RFC 3261
CANCEL	Cancel a pending request	RFC 3261
BYE	Terminate an existing SIP session	RFC 3261
REGISTER	Temporarily bind a device URI to an AOR	RFC 3261
SUBSCRIBE	Establish a session to receive future information updates	RFC 3265
NOTIFY	Deliver information after a SUBSCRIBE	RFC 3265
PUBLISH	Upload status information to a server	RFC 3903
REFER	Request another UA to act upon a URI	RFC 3515
MESSAGE	Transport an Instant Message (IM)	RFC 3428
UPDATE	Update session state information	RFC 3311
PRACK	Acknowledge a provisional response	RFC 3262
INFO	Transport mid call signaling info	RFC 2976

A URI can be expressed in two ways:

- Using an address of record (AOR). This requires access to a DNS server to determine the SIP server. For example, sip:johndoe@somewhere.com will need to get resolved to sipserver1.somewhere.com.

- Using a Fully Qualified Domain Name (FQDN). This does not require any DNS resolution. For example sip:johndoe@sipserver1.somewhere.com or equivalently sip:johndoe@68.249.23.100

4.2.2.3 Protocol Details

SIP is a request-response protocol. The request is called a SIP method. There are six basic methods defined in the base SIP RFC. The remaining methods have been extended via other RFCs.

A response is indicated via a code. The first digit denotes the class of the response. There are six classes. The second part gives the reason code (Table 4.3).

Table 4.3 SIP error responses.

Class	Code	Examples
Informational or Provisional	1xx	100 Trying
		180 Ringing
		183 Session Progress
Success	2xx	200 OK
		202 Accepted
Redirection	3xx	300 Moved
		302 Multiple Choices
		305 Use Proxy
Client Error	4xx	401 Unauthorized
		403 Forbidden
		404 Not Found
		415 Unsupported Media Type
		486 Busy Here
		428 Use Identity Header
Server Error	5xx	501 Not Implemented
		503 Service Unavailable
Global Error	6xx	600 Busy Everywhere
		603 Decline

A SIP message exchange follows the path of a trapezoid as shown in Figure 4.5. A UA to UA session setup will be routed through a pair of proxy servers. When a two-way handshake is done, subsequent signaling, including the media session, will take place directly between the two UAs. The proxy will not be involved.

Let's examine a typical SIP message. The message begins with a Start-Line. The Start-Line can either be a Request-Line or a Status-Line. In this case it is a Request and shows that this is an INVITE message to annie@msn.com for a multimedia session based on a SIP v2.0 request. The URI in the Request Line is referred to as the Request-URI.

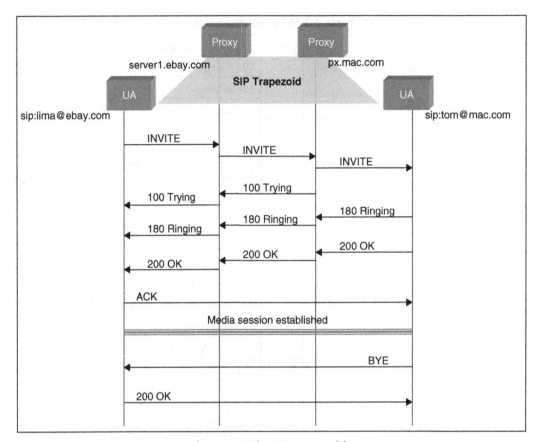

Figure 4.5 The SIP trapezoid.

The Message header shows that the user is expecting the response at 192.168.1.100, port 9816. The maximum number of hops is set to 70 as defined by the *Max-Forwards*. The *Contact* details give a fully qualified description of the owner of the session. The *To* field identifies the message is intended for Annie Frid *From* Charlie Brown. The *Call-ID* is a random number generated for this call. The *CSeq* identifies that in this session, there is one invocation for the INVITE in this dialogue. The *Allow* field identifies all the methods that are permissible for this session. The next few fields describe the nature of the message body. The message body in this case is an SDP payload of size 423 bytes as described by the *Content-Type* and *Content-Length*, respectively, and is generated by a *User-Agent* IMSFone.

	SIP-METHOD	Start Line. Identifies the SIP method, such as INVITE, OPTIONS, and so forth.
Header	Via	Provides the address at which the message originator is expecting to receive the response.
	Max-Forwards	An integer value that contains the maximum number of hops permissible for the message to reach its destination.
	To	Identifies the display name corresponding to the destination party of the message.
	From	Identifies the display name corresponding to the destination party of the message.
	Call-ID	Identifies a unique identifier for the call, which is generated by a random number and the host's IP address.
	CSeq	Contains a command sequence number and a method name. It is used to track new methods invoked in a dialogue.
	Contact	Represents the FQDN of the owner of the session.
	Content-Type	Provides the description of the body of the message.
	Content-Length	Provides the length of the body in bytes.
	Body	Contains the body of the message.

```
Request-Line
  INVITE sip:annie@msn.com SIP/2.0
Message Header
  Via: SIP/2.0/UDP 192.168.0.100:9816;branch=z9hG4bK-d87543-
7a24ff67042ad744-1--d87543-;rport
  Max-Forwards: 70
  Contact: <sip:856634@192.168.0.100:9816>
  To: "Annie Frid"<sip:annie@msn.com>
  From: "Charlie Brown"<sip:856634@svc.alpha.com>;tag=4a21e156
  Call-ID: ZjYzYWMxZDZiMDc4Yjk2OTZlMmYyNjk4ZTZlMjJkMTc.
  CSeq: 1 INVITE
  Allow: INVITE, ACK, CANCEL, OPTIONS, BYE, REFER, NOTIFY,
MESSAGE, SUBSCRIBE, INFO
  Content-Type: application/sdp
  User-Agent: IMSFone release 1011s stamp 41150
  Content-Length: 423
Message body
```

4.2.2.4 SIP Message Headers

We see in Table 4.4 a comprehensive list of the headers that can be used in the IMS SIP messages. Most of the headers are defined in RFC 3261. At the time of writing, there are additional headers for consideration such as Geolocation, which have not been ratified in an IETF RFC at the time of writing. The usage of the headers in terms of being mandatory or optional depends upon the SIP message and is detailed in TS 124.229 for each message type.

Table 4.4 SIP headers.

Header	Defined in	Header	Defined in	Header	Defined in
Accept	RFC 3261	From	RFC 3261	Reply-To	RFC 3261
Accept-Contact	RFC 3261	History-Info	RFC 4244	Request-Disposition	RFC 3841
Accept-Encoding	RFC 3261	In-Reply-To	RFC 3261	Require	RFC 3261
Accept-Language	RFC 3261	Join	RFC 3261	Retry-After	RFC 3261
Alert-Info	RFC 3261	Max-Forwards	RFC 3261	Route	RFC 3261
Allow	RFC 3261	MIME-Version	RFC 3261	Security-Client	RFC 3329
Allow-Events	RFC 3265	Min-Expires	RFC 3261	Security-Server	RFC 3329
Authentication-Info	RFC 3261	Min-SE	RFC 3261	Security-Verify	RFC 3329
Authorization	RFC 3261	Organization	RFC 3261	Session-Expires	RFC 3261
Call-ID	RFC 3261	Path	RFC 3327	Server	RFC 3261
Call-Info	RFC 3261	Priority	RFC 3325	Subject	RFC 3261
Contact	RFC 3261	Privacy	RFC 3261	Supported	RFC 3261
Content-Disposition	RFC 3261	Proxy-Authenticate	RFC 3261	Timestamp	RFC 3261
Content-Encoding	RFC 3261	Proxy-Authorization	RFC 3261	To	RFC 3261
Content-Language	RFC 3261	Proxy-Require	RFC 3261	Unsupported	RFC 3261
Content-Length	RFC 3261	Reason	RFC 3326	User-Agent	RFC 3261
Content-Type	RFC 3261	Record-Route	RFC 3261	Via	RFC 3261
CSeq	RFC 3261	Referred-By	RFC 3892	Warning	RFC 3261
Date	RFC 3261	Reject-Contact	RFC 3261	WWW-Authenticate	RFC 3261
Error-Info	RFC 3261	Replaces	RFC 3261	Extension Headers	RFC 3261
Expires	RFC 3261				

IMS SIP Extension Headers

IMS introduces a set of Private Headers (P-Headers) as a set of extension headers to SIP. These headers are defined in the RFCs described in Table 4.5. At the time of writing there are additional headers for consideration for RFCs such as P-Asserted-Service, P-Preferred-Service, P-Early-Media, and P-Profile-Key.

Table 4.5 IMS SIP extension headers.

P-Headers	Defined in	Purpose
P-Associated-URI	RFC 3455	This extension allows a registrar function to return a set of associated URIs for a registered address-of-record.
P-Called-Party-ID	RFC 3455	This extension header is populated by a proxy with the Request-URI received in the request.
P-Visited-Network-ID	RFC 3455	This extension is used to convey the address of the visited network to the proxy or the registrar in the home network.
P-Access-Network-Info	RFC 3455	This extension contains the information on the access network being used by UE for IP connectivity. The information is intended for the first proxy (P-CSCF) and is stripped by the message before forwarding the message.
P-Charging-Function-Addresses	RFC 3455	This extension is required to populate the addresses of the charging functions; the ECF and the CCF.
P-Charging-Vector	RFC 3455	This header contains the ICID and the IOI information for the charging information.
P-Asserted-Identity	RFC 3325	The asserted-identity is established by the proxy server upon authenticating the originating user. It establishes a trust relationship.
P-Preferred-Identity	RFC 3325	The UE provides its preferred identity to the proxy server that it wishes to be used for the asserted-identity, after the trust relationship is established.
P-Media-Authorization	RFC 3313	This extension header contains tokens that are required for authorizing QoS for media streams, applicable to the Rel5/6 policy architecture.
P-User-Database	RFC 4457	This header carries the address of the HSS and is exchanged between the P-CSCF and I-CSCF.

4.2.3 SigComp

The elegance of text-based protocols is blemished by their large packet size for transmitting over a wireless media. The traditional Global System for Mobile (GSM) and Code Division Multiple Access (CDMA) models have used efficient message formats for wireless transmission. While the third generation (3G) wireless speeds have increased tenfold to greater than a 1Mb/sec in contrast to 100Kbps from 2G/2.5G, large packet sizes result in the following challenges:

- Increased Call Setup time. SIP-based General Packet Radio Service (GPRS)/ Enhanced Data Rates for GSM Evolution (EDGE) systems have been shown to take nearly double the setup time in GSM.

- The Store and Forward behavior of SIP results in latency and long roundtrip delays.

- Fitting large signaling packets into a permissible number of Radio Network frames becomes constrained.

An average size of an IMS packet for a Push to Talk over Cellular application is in the range of 1 to 1.5Kbytes, in comparison to a GSM Um packet of about 120 bytes. Reducing the size of this packet can be achieved by a suitable Signaling Compression (SigComp) technique. This technique is based on RFC 3320 and further extended by RFC 3485 for SIP/SDP applicability. SigComp provides a framework, which is compression algorithm independent. It is designed to work with pre-arranged parameter agreements, without requiring any application level negotiation. It functions on a single message protocol with an optional feedback mechanism.

The SigComp implementation has the capability to both compress and decompress a message (Figure 4.6). The application provides an application message and a compartment identifier to the compression dispatcher. The compression dispatcher invokes a particular compressor function on a per-compartment basis, using the compartment identifier provided by the application. The compressor chooses the algorithm for compression and provides the compressed message to the compression dispatcher, which in turns provides it to the transport for the endpoint. The decompression process is similar, where the decompression dispatcher receives a compressed message from the remote endpoint. It selects a particular instance of the decompressor function that uses a Universal Decompressor Virtual Machine (UDVM), which decodes the

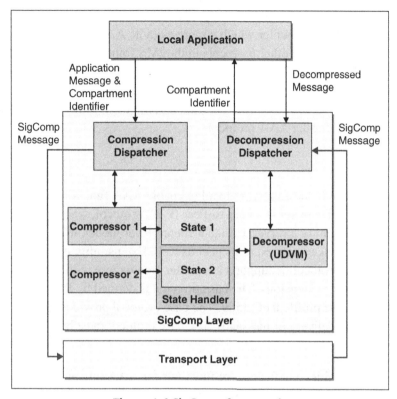

Figure 4.6 SigComp framework.

received message. It provides a compartment identifier to the Decompression dispatcher, if the application wants to access an existing state or create a new state.

The UVDM uses a pre-defined dictionary of SIP/SDP phrases, which are defined in RFC 3485.

Compression techniques in the IMS for SIP are applied at the UE and the edge of the IMS network, which is also the first SIP proxy function.

4.3 AAA Protocol—Diameter

The Remote Authorization for Dial-in User Service (RADIUS) protocol has been incumbent in the IP world to provide the function of Authorization, Authentication, and Accounting (AAA). Designed initially for dialup services, RADIUS has been well deployed with Mobile-IP and CDMA systems. There are, however, a number of weaknesses in RADIUS, to counter which a more suitable AAA protocol was required for IP multimedia. Unlike the other protocols in the IMS framework that we observed, such as SIP and SDP, RADIUS lacks the prime feature of not being an extensible protocol. While initial IMS deployments may encounter the need to interoperate with legacy RADIUS AAA servers, the third-generation partnership project (3GPP) selected Diameter as its choice for the AAA protocol.

4.3.1 Diameter

Diameter is an extensible messaging protocol to enable AAA function in IP and multimedia networks. Diameter is a Peer-to-Peer (P2P) protocol, where a peer can function either as a client or a server. Unlike a client-server model, in Diameter, a server has the capability to make a request to its peer as well. Extensibility is an important trait of the Diameter protocol. It supports a modular architecture with the base protocol and application-specific extensions. The base protocol is defined by RFC 3588 and a corresponding transport profile RFC 3589. The base protocol provides support for the reliable transport and delivery of messages. The base protocol must be used along with an application extension.

The base protocol provides a default set of functions for AAA functions. A AAA application extension can choose to rely on any or none of these functions (implemented as state machines) to provide application-specific value. For instance, Diameter Credit Control Application (DCCA RFC 4006) utilizes the authorization and authentication part of the base protocol, but implements its own set of accounting state machines to support the online charging function for the application. In the initial IMS Rel5 and Rel6 definitions, only the application extensions for each of the IMS interfaces to access subscriber, policy, and charging data were defined. Rel7 has also introduced the access for PWLAN and other broadband access. This has added newer interface support and the NASREQ and EAP modules for authorization and authentication, respectively. Ongoing work with Mobile IPv4 and MIPv6 may be augmented to the IMS standards at some point. This is shown in Figure 4.7.

Diameter can work over TCP, SCTP, or UDP. 3GPP has, however, recommended a secure and reliable transport to make it a suitable choice for charging and authorization. SCTP is recommended more than TCP for its reliability in the field.

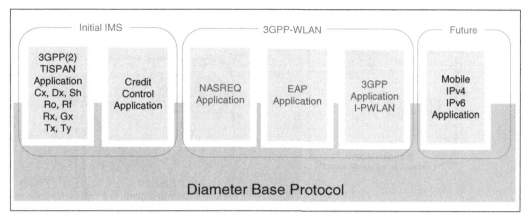

Figure 4.7 Diameter extensible protocol.

Client	Network Edge Device Performing Access Control; eg, NAS, Foreign Agent.
Server	Controlling Entity of AAA functions for a particular domain; eg, HSS.
Relay Agent	Routes Diameter messages within known peers in supported realms. May modify routing information (only).
Proxy Agent	Also routes messages, but can modify message content to enable policy, resource usage, admission and provisioning.
Redirect Agent	Enables routing to other domains within roaming agreements by notifying the requesting peer with the routing information.
Translation Agent	Protocol translation function such as RADIUS-Diameter conversion.

Figure 4.8 Diameter entities.

Diameter, like SIP, has several functional elements as shown in Figure 4.8. These comprise a client, server, and agents. A Diameter session is a request-response interaction, and in the simplest case this can be achieved by a client-server configuration. Agents are useful for relaying, redirecting, translating, and proxy of requests.

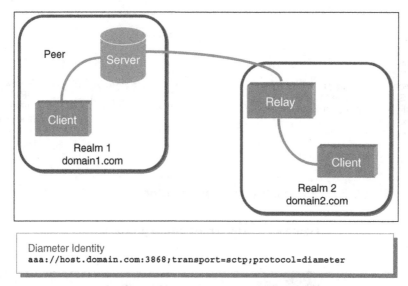

Figure 4.9 Diameter entity relationship.

Peers can be statically configured or dynamically discovered. In most deployments today, these are set for static configuration. Two diameter entities share a peer relationship, if they reside within the same realm. In diameter parlance, a realm is the equivalent of a domain. A relationship with an entity in a different realm can be established by going through a hop with assistance from a relay or proxy agent. This is illustrated in Figure 4.9.

Following a process of discovering the peer, which as we saw is mostly done through static configuration, the initial handshake between two peers is established via capabilities exchange messages as we observe in Figure 4.10. This message exchange of a capabilities request and answer establishes the basic capability set, which both entities negotiate. In this exchange, each peer provides the set of application identifiers it can support. The peers can agree on a common set of applications, which will be used in the interaction. The exchange also allows the peers to agree on common security mechanisms. Following the capabilities exchange, the two peers are ready to communicate and can set up a session, either between themselves or toward another entity, using the peer as a relay or proxy.

Diameter provides a fault-detection mechanism with heartbeats. Heartbeats are exchanged for transport failure detection. These Device Watchdog messages are exchanged, if there is no exchange of AAA messages. Failover/failback mechanisms are invoked when transport failures are detected. An alternate peer is selected for all pending and new requests.

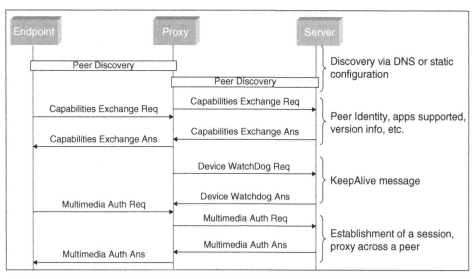

Figure 4.10 Initial message exchange.

Diameter commands are expressed as request/answer pairs. Each pair is identified with a command code (Table 4.6).

A Diameter message, as we see in Figure 4.11, comprises a fixed length header part and a set of Attribute Value Pairs (AVP). The header contains protocol-, command-, and session-related information.

The diameter protocol version is defined in the Protocol field. The command, the length of the message, and other message qualifiers are described in the CommandCode, MessageLength, and Flags fields, respectively. For extensibility, vendor-specific commands can also be added, so each command is associated with a Vendor ID. The Application ID is an important field, which is used to signify which Diameter application the message is intended for. The Base protocol can support multiple Diameter applications. The Application ID enables the base protocol stack to route the message to the Accounting, Authentication, or the other Diameter applications. Session-related information is described by two identifiers. The Hop-by-Hop Identifier correlates the information with the peer (next-hop or neighbor) entity. The end-to-end identifier correlates the information with the endpoint for the transaction, which may be several hops away.

Table 4.6 Diameter commands.

Command	Abbr	Code	Description
Capabilities Exchange-Request	CER	257	Request a peer for its identity, capabilities, diameter applications supported, and security mechanisms
Capabilities-Request-Answer	CRA	257	Response from the peer to a CER providing its identity, capabilities, Diameter applications supported and security mechanisms
Device-Watchdog-Request	DWR	280	This request is sent to the peer when there has not been any exchange of traffic in a configured time interval
Device-Watchdog-Answer	DWA	280	The answer to the watchdog request
Accounting-Request	ACR	271	Request accounting information from a peer
Accounting-Answer	ACA	271	Response with the accounting information
Disconnect-Peer-Request	DPR	282	Request sent to a peer to bring down the transport connection
Disconnect-Peer-Answer	DPA	282	Acknowledgement sent to the peer for bringing down the transport connection
Re-Auth-Request	RAR	258	A request sent by a Diameter server to a client to re-authenticate the device
Re-Auth-Answer	RAA	258	Answer sent by the client in response to the re-authentication request
Abort-Session-Request	ASR	274	Request sent by a server to a client to stop a particular session
Abort-Session-Answer	ASA	274	Answer sent by the client in response to an abort session request
Session-Termination-Request	STR	275	The request is sent by the client to the server indicating that the authenticated/authorized session is being terminated
Session-Termination-Answer	STA	275	The answer from the server to acknowledge the session termination

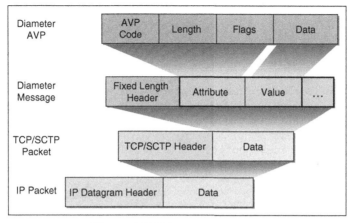

Figure 4.11 Diameter message.

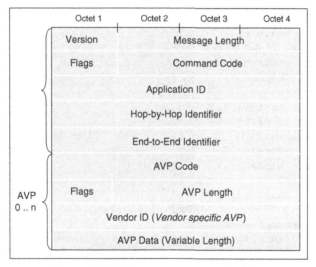

Figure 4.12 Diameter message details.

The Diameter protocol allows grouping of AVP values. This implies that the Data field is actually a sequence of AVPs in a Grouped AVP. Nesting of Grouped AVPs is allowed. It is possible to include an AVP with a Grouped type within a Grouped type.

The following is a sample Diameter message corresponding to the CER. Diameter is not expressed in the same ASCII format as SIP or SDP as we saw earlier. This is a decoded version. Figure 4.12 provides the Diameter message details.

```
Diameter Protocol
  Version: 0x01
  Length: 184
  Flags: 0x80 (Request)
  Command Code: Capabilities-Exchange-Request
  ApplicationId: Diameter Common Messages (0)
  Hop-by-Hop Identifier: 0xa003cfb6
  End-to-End Identifier: 0xb1b1fb1c
  Attribute Value Pairs
    Origin-Host (DiameterIdentity) l:0x18 (24 bytes) (24 padded bytes)
      AVP Code: Origin-Host (264)
      Flags: 0x40 (Mandatory)
      AVP Length: 24
      Identity: srv.test.ims.com
    Origin-Realm (UTF8String) l:0xf (15 bytes) (16 padded bytes)
      AVP Code: Origin-Realm (296)
      Flags: 0x40 (Mandatory)
      AVP Length: 15
      UTF8String: ims.com
    Host-IP-Address (IpAddress) l:0xe (14 bytes) (16 padded bytes)
      AVP Code: Host-IP-Address (257)
      Flags: 0x40 (Mandatory)
      AVP Length: 14
      Address Family: IPv4 (1)
      IPv4 Address: 192.168.56.77
    Vendor-Id (Enumerated) l:0xc (12 bytes) (12 padded bytes)
      AVP Code: Vendor-Id (266)
      Flags: 0x40 (Mandatory)
      AVP Length: 12
      Value: 0x00000002 (2): IMS
    Product-Name (UTF8String) l:0x28 (40 bytes) (40 padded bytes)
      AVP Code: Product-Name (269)
      Flags: 0x00 (<None>)
      AVP Length: 40
      UTF8String: IMS TestSuite Diameter Component
    Supported-Vendor-Id (Unsigned32) l:0xc (12 bytes) (12 padded bytes)
      AVP Code: Supported-Vendor-Id (265)
      Flags: 0x40 (Mandatory)
      AVP Length: 12
      Value: 0x000028af (10415)
    Inband-Security-Id (Enumerated) l:0xc (12 bytes) (12 padded bytes)
      AVP Code: Inband-Security-Id (299)
      Flags: 0x40 (Mandatory)
```

```
     AVP Length: 12
     Value: 0x00000000 (0): NO_INBAND_SECURITY
   Vendor-Specific-Application-Id (Grouped) 1:0x20 (32 bytes) (32
padded bytes)
     AVP Code: Vendor-Specific-Application-Id (260)
     Flags: 0x40 (Mandatory)
     AVP Length: 32
     Vendor-Specific-Application-Id Grouped AVPs
       Vendor-Id (Enumerated) 1:0xc (12 bytes) (12 padded bytes)
         AVP Code: Vendor-Id (266)
         Flags: 0x40 (Mandatory)
         AVP Length: 12
         Value: 0x000028af (10415): 3GPP
       Auth-Application-Id (Unsigned32) 1:0xc (12 bytes) (12 padded
bytes)
         AVP Code: Auth-Application-Id (258)
         Flags: 0x40 (Mandatory)
         AVP Length: 12
         Value: 0x01000001 (16777217): 3GPP Sh
```

4.3.2 COPS

The Common Open Policy Service (COPS) provided the policy framework for the initial IMS releases. With release 7 onward, the use of Diameter for the policy interfaces obviates the need for COPS. COPS has also been deployed with cable networks for bandwidth control. We will only briefly cover this protocol. The base protocol for COPS is defined in RFC 2784. The protocol defines the exchange for policy-related information between a Policy Decision Point (PDP) and the Policy Enforcement Point (PEP). COPS focuses on the communication between the two entities. It is agnostic of the semantics of the data itself, which is exchanged. COPS supports different clients corresponding to different policy management areas (for example, Policy Provisioning (PR), Differentiated Services [Diffserv], or Resource ReSerVation Protocol [RSVP]).

4.4 Media Protocols

Transporting multimedia content in IP packets exposes some challenges that were not seen while dealing with session protocols. The digitization of audio and video is done by sampling the signals at specific time intervals. These samples are packetized and sent for transmission. If the playback of the received packets is not played back continuously, the multimedia content will exhibit delays and jitter to the human eyes and ears. The unpredictable arrival of IP datagrams manifests as these delays, and jitter while carrying media-related payload. Congestion in the network can also contribute to the packets not arriving in real-time. For a signaling or a session protocol, this could be experienced as a delay, but for a media protocol it results in obsolete packet delivery. Further, multimedia traffic by its nature tends to occur in bursts. This requires the receiver to be able to buffer and smooth the data. A buffer overrun will result in packet loss, and a buffer under-run will result in gaps in the playback. It is therefore essential for a media protocol to support the capability to handle real-time traffic, enable sequencing, and provide a control of the packet delivery.

4.4.1 RTP

The Real-Time Protocol (RTP) provides mechanisms for an end-to-end transport for multimedia data over an IP packet network. It supports the essential support to provide timing and sequencing in data transmission and playback for multimedia applications. RTP by itself is only a packet protocol. It does not guarantee sequenced delivery or timeliness of packets themselves. It provides the capability to the application to adapt to the information as provided by RTP. RTP also works in conjunction with the Real-Time Control Protocol (RTCP). RTCP aims to provide control packets that enable feedback on the delivery of the packets.

RTP uses User Datagram Protocol (UDP) for the transport. As shown in Figure 4.13 the RTP header and the payload, which is the encoded media, are encapsulated in the UDP data. We examine the RTP header to understand how it supports the requirements for multimedia traffic.

The first octet in the RTP header contains a 2-bit RTP version identifier V. The third bit P contains a padding indicator to denote if padding octets are used at the end of the header. The fourth bit X denotes an extension header, if it is set. The RTP header contains a variable number of Contributing Source (CSRC) Identifiers, between 0 to 15. The CSRC count gives the number of these identifiers.

The second octet contains a 1-bit marker M, the interpretation of which is defined by a profile. It is intended to allow significant events such as frame boundaries to be marked in

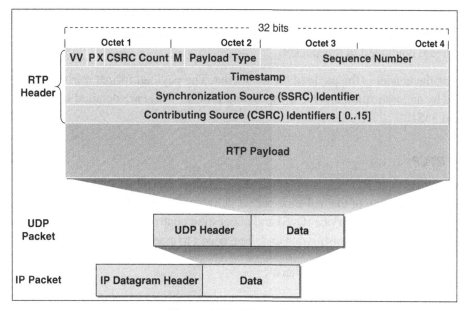

Figure 4.13 RTP packet.

the packet stream. This is followed by a 7-bit payload type, which identifies the format of the RTP payload and determines its interpretation by the application. A profile specifies a default static mapping of payload type codes to payload formats. The definition of RTP payloads is kept flexible, to accommodate new payload types.

The next two bytes are used for a sequence number. This provides a mechanism to ensure sequencing and may be used by the receiver to detect packet loss and to restore packet sequence. This field increments by one for each RTP data packet sent.

The next 4-octet field Timestamp assists in the real-time aspect of the protocol. This stores the sampling instant of the first octet in the RTP data packet. The sampling instant must be derived from a clock that increments monotonically and linearly in time to allow synchronization and jitter calculations.

The next 4-octet field Synchronization source (SSRC) is used to identify the source of the RTP stream. This identifier is chosen randomly, with the intent that no two synchronization sources within the same RTP session will have the same SSRC identifier.

The variable part of the RTP header contains a list of Contributing source identifiers (CSRC). This identifies the contributing sources for the payload corresponding to a mixed media stream contained in this packet.

Following the header is the payload for the media. The payload format must be specified by an application and a corresponding profile. These are detailed in RFCs 3550 and 3551.

4.4.1.1 RTCP

The participants in an RTP session periodically exchange control packets to provide feedback on the quality of the real-time packet delivery. These RTCP packets provide a monitoring of Quality of Service (QoS) and congestion control. This information is valuable to the sender to adjust its transmission and the receiver to assess congestion handling.

There are five types of Control packets.

- **RR: Receiver Report** Receiver reports are generated by participants that are not active senders. They contain reception quality feedback about data delivery, including the highest packets number received, the number of packets lost, inter-arrival jitter, and timestamps to calculate the round-trip delay between the sender and the receiver.

- **SR: Sender Report** Sender reports are generated by active senders. In addition to the reception quality feedback as in RR, they contain a sender information section, providing information on inter-media synchronization, cumulative packet counters, and number of bytes sent.

- **SDES: Source Description Items** These contain information to describe the sources.

- **BYE** Indicates end of participation in the RTP session.

- **APP** Application-specific functions. This is being pursued for experimental use.

Unlike usual data transmission, RTP does not offer any form of reliability or flow/congestion control. It provides timestamps, sequence numbers as hooks for adding reliability, and flow/congestion control. The implementation is application-dependent.

RTP is a flexible framework protocol. It is open to new payload formats and new multimedia applications. By adding new profile and payload format specifications, one can tailor RTP to new data formats and new applications.

RTP/RTCP provides the functionality and control mechanisms necessary for carrying real-time content. However, RTP/RTCP leaves the assembly and synchronization to the application level.

The flow and congestion control information of RTP is provided by RTCP sender and receiver reports.

4.4.1.2 In Perspective with SIP/SDP

A common misconception prevails with the control moniker in the RTCP. RTCP is not the control protocol for an RTP session. Its objective, as we saw earlier, is to provide QoS feedback and flow/congestion control information about RTP sessions. So, how does the RTP session get established? It is by the definition in the SDP and exercised by the offer-answer model.

The Media Line (m=), as we saw in Table 4.1, contains the information about the media session. This has four parameters:

- **The Type of Media** Audio, video.

- **Port** This is the IP port number that will be used for the multimedia session. This must be an even number for an RTP session.

- **Transport** This defines the protocol. RTP/AVP connotes the Audio Video profile with RTP.

- **Format-List** Payload types defined in the RTP/AVP.

With an RTP media session, SDP can also use a special media attribute rtpmap to bind a codec to the payload type. As we observed in the RTP payload type, a static definition is available by the profile. The rtpmap attribute provides a dynamic definition.

Consider the following as an example.

```
m=audio 45732 RTP/AVP 119 0 8
a=rtpmap:119 BV32-FEC/16000
```

The Originator of the session is offering in the SDP definition via a SIP message such as an INVITE, to establish an audio stream on port 4572 using the RTP audio video profile. It supports the following media formats:

- 0—ITU-T G.711 PCMU codec

- 8—ITU-T G.711 PCMA codec

- 119 is defined as a payload type that can support the BV32-FEC encoding at a clock rate of 16000

Once the offer is accepted, which is when an Acknowledgement (ACK) is received, the two endpoints will commence a media session on the assigned port.

4.4.2 MEGACO/H.248

The Media Gateway Control Protocol (Megaco)/H.248 is a general-purpose gateway control protocol that has been standardized by the cooperative effort of the IETF and the Telecommunication Standardization Sector (ITU-T). Designed for master-slave paradigm for a media gateway controller to control the media gateway, it finds similar applicability in IMS for the Media Resource Function Controller (MRFC) to control the Media Resource Function Platform (MRFP). We examine the connection model and the structure of the protocol. H.248 proposes a logical connection model based on abstract description. There are two abstractions: a *termination* and a *context*.

A termination is a logical entity in a media gateway comprising one or more media streams. A stream carries a media type, such as RTP or a Circuit Switched (CS) bearer. Each termination has a set of properties. These properties are associated with descriptors that can be used in the commands to direct them.

A collection of terminations is called a context. Terminations within a context have the capability to route media flows between themselves. This is shown by the * box in Figure 4.14. A context is created when a termination is added to it and is destroyed when the last termination is disassociated. A context with a single termination is called a null context. Both termination and context have their unique identifiers.

An MGC or the MRFC communicates a command set via a hierarchically nested structure in a Megaco/H.248 message, as shown in Figure 4.15. Each message comprises a header and a set of transactions. A transaction may be a request or a reply and include a set of actions to be performed. The transaction set is ordered and is

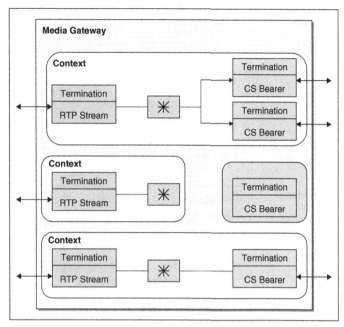

Figure 4.14 H.248 model.

executed in sequence. Each action contains one or more commands, each of which relates to a specific context.

Commands relate to the operations that can be performed on the connection model we saw before. These are essentially to manipulate and control the properties of the terminations and the contexts. The parameters in these commands are termed as descriptors. The following is the command set of Megaco/H.248.

Megaco is expressed as a text-based protocol. It has a binary version, which is encoded in ASN.1 as well. The following is an example of a Megaco message to add an RTP stream to a context.

```
MEGACO/1 [192.168.1.100]:2944 Transaction = 10006{Context=${Add
= 0MRFP1,Add = ${Media{Stream = 1{LocalControl{Mode =
ReceiveOnly},Local{
v=0
c=IN IP4 $
m=audio $ RTP/AVP 0
}}}}}}
```

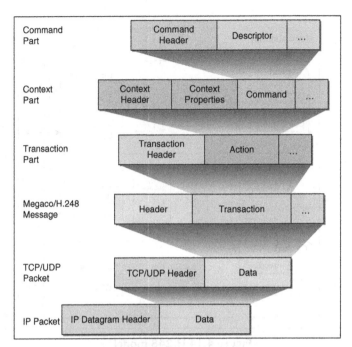

Figure 4.15 Megaco message.

Table 4.7 Megaco/H.248 commands.

Command	Description	Direction
ADD	Adds a termination to a context. This also implicitly creates a context when the first transaction is added.	MGC→MG
MODIFY	Alters the behavior of an existing termination.	MGC→MG
SUBTRACT	Removes a termination from a context. This also implicitly deletes a context when the last termination is removed.	MGC→MG
MOVE	Moves a termination from one context to another.	MGC→MG
AUDITVALUE	Determines characteristics of a termination or the complete MG/MRFP.	MGC→MG
AUDITCAPABILITY	Determines capabilities of a termination or the complete MG/MRFP.	MGC→MG
NOTIFY	Sets the MG to report an occurrence of an event or a set of events.	MGC←MG
SERVICECHANGE	Notifies the responder of a new service change.	MGC↔MG

4.5 IP Access and Addressing

4.5.1 DNS

The DNS in IP networks uses a hierarchical name structure (organized as a tree). This scheme of addressing provides addressing by named resources instead of physical resources. Domain named addressing in a familiar format such as www.ims-world.com is easier than a physical address such as 72.184.23.1. In IMS we need to resolve named addressing in two different situations. Initially, for a device to obtain access, it needs to request entities known to it by a name. Subsequently, most SIP and Diameter interaction requires the resolution of the FQDN. Let's understand how this applies. DNS as defined by RFC 1034 comprises of three elements.

A *DNS server*, which organizes data for each domain it supports, and describes the global properties of the domain and its hosts (or services). This data is defined in the form of textual Resource Records (RR) organized in Zone files. The format of Zone files is defined in RFC 1035.

The *Name Server* program, which responds to queries from local or remote hosts. It will identify the zones it is responsible for and determine any caching behavior for the name resolution.

Finally, the *Resolver*, located on each host, provides a means of translating a user request for name resolution into one or more queries to DNS servers using UDP (or TCP) protocols.

A subset of record type formats of interest to us in the IMS world, maintained in a DNS server are the following (Table 4.8):

Table 4.8 DNS record type formats.

RR	Description
A	IPv4 Address record. An IPv4 address for a host.
AAAA	IPv6 Address record. An IPv6 address for a host.
SRV	The Service Record defines services available in zone; e.g., SIP, HTTP, and so forth.
NAPTR	Naming Authority Pointer Record. General-purpose definition of a rule set to be used by applications; e.g., SIP applications.

In most of the initial UE access to the network, A and AAAA record formats help in the name resolution of the access gateway. SIP name resolution requires either the SRV or NAPTR format. For Service Record (SRV) records, the DNS server will need to be set up with entries in the following format (specified in RFC2782):

_Service._Proto.Name TTL Class SRV Priority Weight Port Target

For example, the entry to resolve sip:cbrown@sipserver.ims-domain.com

_sip._udp.ims-domain.com 86400 IN SRV 10 10 5060 sipserver.ims-domain.com.

NAPTR NAPTR records (specified in RFC3403) are in the form of

"domain-name TTL Class NAPTR order preference flags service regexp target"

For example, the entry to resolve sip:cbrown@sipserver.ims-domain.com

"ims-domain.com IN NAPTR 60 50 "s" "SIP + D2U" "" _sip._udp.ims-domain.com."

SIP + D2U connotes UDP for the transport.

To understand the interaction with these RR records, a SIP UA first sends a DNS NAPTR request to the domain specified in the Request-URI. If valid records are returned, an appropriate transport is identified. If no NAPTR records are returned, a DNS SRV request is sent based on the transport preferred by the User Agent Client (UAC). If valid records are returned, the request is sent to the preferred proxy. As a last resort, if no SRV records are found, a DNS A record request is sent for the domain in the request URI. In the case of sip:cbrown@sipserver.ims-domain.com the request would be for the IP address of ims-domain.com. If a valid IP address is returned, the request is sent to that address using UDP.

4.5.2 DHCP

There are two methods for an IP endpoint to join a network and have a recognizable IP address. Its address can either be configured statically, or it can be assigned by a server on the network supporting the Dynamic Host Configuration Protocol (DHCP). DHCP is one of the mechanisms for a UE to obtain an IP address to join the IMS network. DHCP is carried in the IP Bootstrap Protocol (BOOTP) packets.

An IP endpoint that connects to the network, broadcasts for a DHCP server on the network. It does so by sending a DHCP discover message on UDP port 67 (Figure 4.16). At this stage, its IP address is unknown, so it uses 0.0.0.0 for its own source identifier

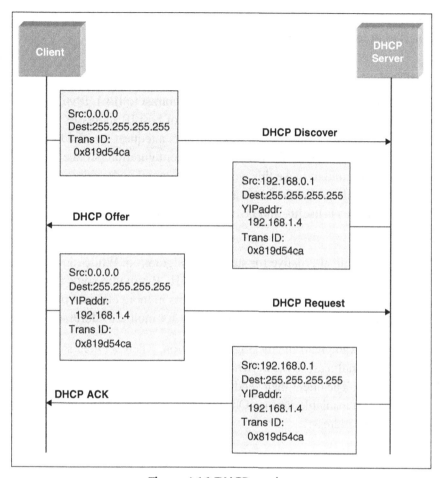

Figure 4.16 DHCP session.

and a broadcast mask of 255.255.255.255 for the destination. It also sends a transaction identifier, which the DHCP server can use to correlate this request. A DHCP server, 192.168.0.1 as shown as an example, upon receiving this request will respond to the client with a DHCP Offer for an IP address. This will contain your (client) IP address and a lease time, the duration for which this IP address is valid. The client now responds with a DHCP Request for this IP address. The DHCP server acknowledges this request and responds with a DHCP acknowledgement (ACK).

A DHCP offers an IP address on a lease to a client. Upon expiry of the lease, the address must be renewed or returned to the pool.

4.5.2.1 DHCPv6

The allocation of IPv6 addresses has two options in contrast to IPv4. IPv6, by its nature, supports a mechanism for stateless autoconfiguration of host IP addresses. When an IP endpoint is first connected to the network, it broadcasts a request for its address. In this case, the routers can respond with the network layer configuration parameters. Routers must be suitably configured to make this work. The endpoint adds additional information to generate a unique address. While this is a suitable approach, having control over a managed set of addresses is useful, and providing adequate network information to the routers is not easy.

DHCPv6 thus provides an alternative for stateful configuration. While most of the principles are similar to DHCP for IPv4 addresses, DHCP provides some advantages. It does not work with the BOOTP protocol, which results in more efficient packets and a leaner set of options. The DHCPv6 client can request for multiple addresses, not just one.

Figure 4.17 gives an example of the message interaction, where a client solicits its address by sending a multicast to the network. In this case, two servers advertise their network configuration. The client then sends a request to one of the servers, and upon receiving a reply has it available for use. Once the lease expires, it will send a request to renew the lease of the address.

4.5.3 GPRS/EDGE

An IMS-enabled device obtains access to the wireless network by establishing a data session with the support of the packet-switched capability of the 2.5G/3G protocols. The 2.5G protocols are GPRS-EDGE/CDMA2000 1xRTT-1xEV and the 3G protocols such as UMTS/CDMA2000 3x. As we see in Table 4.9, these evolutionary protocols provide the transport capable of uplink (sending) speeds ranging from 80 Kpbs to 500 Kbps and downlink (receiving) speeds from 140 Kbps to 500 + Kbps.

The radio capabilities of Universal Mobile Telecommunication System (UMTS) and High Speed Packet Access (HSPA) are significantly different from GPRS to enable higher data rates and low latency. From an IP system perspective, there is not much of a difference.

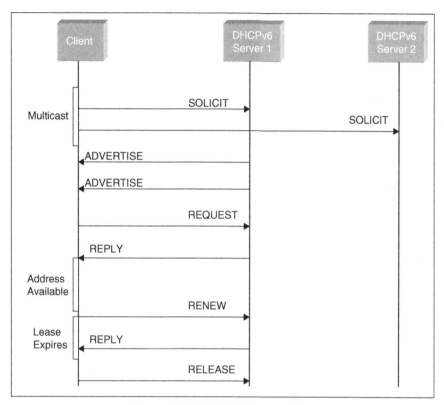

Figure 4.17 DHCPv6 interaction.

Table 4.9 Wireless transports.

Technology	Uplink	Downlink	Latency
EDGE	80 Kbps	140 Kbps	250–300 ms
UMTS	250 Kbps	400 Kbps	150–200 ms
HSDPA + HSUPA	5.76 Mbps	14.4 Mbps	100–125 ms
HSPA +	11 Mbps	42 Mbps	50–75 ms
CDMA 2000 1xRTT	80 Kbps	150 Kbps	250 ms
EVDO Rel0	250 Kbps	500 Kbps	125 ms
EVDO RevA	500 Kbps	800 Kbps	75–100 ms

So we explore the GPRS model of establishing a data session. While we may refer to it as GPRS/EDGE, the IP model applies to UMTS/HSPA as well.

The IP Connectivity to an IMS UE is provided by the PS-Domain protocol by establishing a data session. This model is similar to what the Blackberry and iPhone use today. In GPRS/EDGE, the data session is established by a Packet Data Protocol (PDP) context. The UE requests the GPRS entities to assign an IP address and establish a logical connection to exchange IP packets. Once the PDP context is activated, the IP packets can be exchanged.

An MNO uses the concept of an Access Point Name (APN) as the identifier of the logical entity providing packet data service to the subscriber. An APN's role is to provide a range of IP addresses to a group of users for a particular set of data services. An APN, thus seen as a user group, can enable more than one service for an APN as the operator chooses to configure it.

In a GPRS network as illustrated in Figue 4.18, the APN is implemented as the GPRS Gateway Service Node (GGSN), which provides the service access to the P-CSCF. The Serving GPRS Service Node (SGSN) enables the setup of the PDP session with the GPRS. Let's examine how this works.

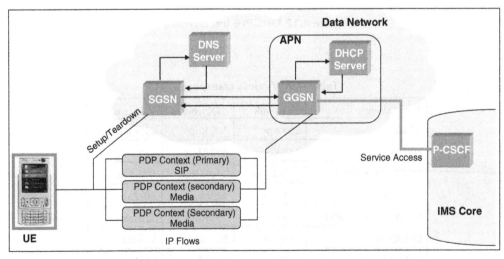

Figure 4.18 The GPRS access architecture.

The UE initiates the PDP context activation procedure (Figure 4.19) to obtain an IP address for the device. The AP specified by the service provider is passed as a parameter. The SGSN initiates a DNS query to find the GGSN corresponding to the APN specified by the mobile. The DNS server provides the GGSN IP address. The SGSN then routes the PDP context activation request to the GGSN corresponding to the APN. The GGSN authenticates the subscription with a RADIUS server. The GGSN now requests a DHCP server for a dynamic IP address for the requesting UE. The DNS server provides the IP address. The GGSN responds back to the SGSN, indicating completion of the PDP context activation procedure. The SGSN replies back to the UE. This signals completion of the PDP context activation.

The underlying protocols that enable this interaction shown in Figure 4.20 are integral in carrying the IMS SIP signaling between the UE and the core network. These are transparent to IMS due to its agnostic nature of the access network. As we saw earlier,

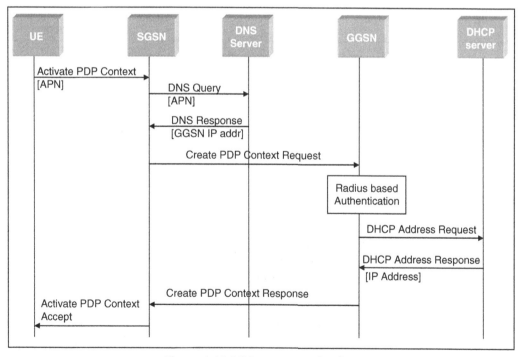

Figure 4.19 PDP context activation.

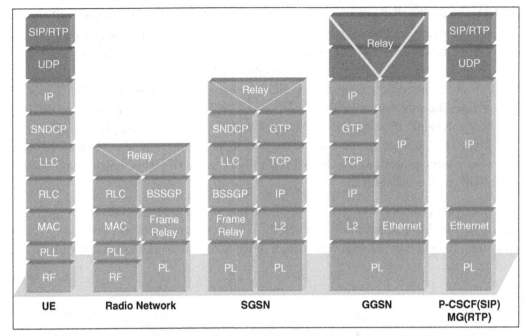

Figure 4.20 Signaling over the GPRS stacks.

the SGSN enables the UE to establish an IP connection with the GGSN, and the SIP/RTP signaling between the UE and the GGSN can commence on the establishment of the PDP context activation. The IP packets are carried over a set of protocols providing a reliable mechanism over the RF interface. The Sub Network Dependent Convergence Protocol (SNDCP) is used between the SGSN and UE to convert network layer IP Protocol Data Units (PDUs) into a suitable format for the underlying network architecture. The Logical Link Control (LLC) is used to provide a ciphered logical link between SGSN and the UE. The Radio Link Control (RLC) is responsible transfer, segmentation, and re-assembly of the PDUs. The Medium Access Control (MAC) controls the access signaling across the air interface. These protocols are correspondingly inter-worked to provide the terrestrial link connection to the SGSN, with a Frame Relay or equivalent protocol.

The IP packets received at the SGSN are encapsulated within a GPRS Tunneling Protocol (GTP), which itself uses TCP for a reliable connection to the GGSN. GGSN to the IP core network uses the standard IP stack.

4.5.4 CDMA2000/1xEV

CDMA provides a comparable mechanism to GPRS/EDGE for the IMS UE to obtain IP connectivity. The protocols and methods to set up the data session, however, have differences. The data session is established by the Point-to-Point Protocol (PPP). The UE requests the CDMA entities to assign the IP address and establish the logical connection for IP packet exchange. Once the PPP session is set up, the IP packets can be exchanged. In the CDMA network, the Packet Data Serving Node (PDSN) provides the service access to the P-CSCF, as shown in Figure 4.21. The PDSN also enables the setup of the PPP session with the UE.

CDMA networks utilize the simple IP or Mobile IP Protocol (MIP) for granting IP connectivity. Mobile IP is preferred as it can support seamless handovers. With simple IP

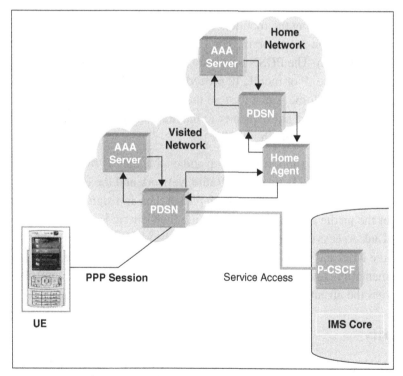

Figure 4.21 The CDMA access architecture.

protocol, a wireless device needs to obtain a new IP address whenever it changes its point of connection. Since the PDSN is responsible for assigning the IP address, moving from PDSN to another PDSN constitutes a change in packet data session. The IP address will have to be re-assigned. A packet data session and a PPP session are concurrent in simple IP.

With mobile IP protocol, the wireless device can maintain the same IP address as it moves between different PDSNs. The packet data session can continue to exist even through multiple PPP sessions. The mobile IP allows the PPP session to be terminated and reestablished without the need to terminate a packet data session. A packet data session can span several PDSNs. This flexibility to retain the IP address spanning across multiple PPP sessions is aided by the Home Agent (HA). The HA is an IP router function that maintains the IP address correlation with the PDSN (also functioning as a Foreign Agent). It can route the messages to the right PDSN that is currently serving the device.

Let's examine how this works. The UE initiates the PPP session setup to obtain an IP address for the device. The request reaches the PDSN of the visited network (or the home network, if it is serving the mobile). The PDSN requests authentication of the mobile by a request to the AAA server. The UE then requests an IP address, following successful authentication. The PDSN determines the nature of the service required for the device to perform simple IP or mobile IP procedures. For Mobile IP support, the PDSN communicates with the Home Agent to store the IP address and routing information. This information can then be used subsequently to maintain the packet data session, when the UE hands over the session to another PDSN.

The underlying stacks that support this setup and enable the SIP/RTP session are shown in Figure 4.22. The protocol between the Radio Network and the PDSN is referred to as the R-P protocol, and is used to transfer both packet data and signaling messages. The packet data or the media flows are tunneled using the Generic Routing Encapsulation (GRE). The Link Access Control (LAC) layer and the MAC layers perform similar functions to the RLC and MAC in the GPRS/EDGE network. The LAC is responsible for transfer, segmentation, and re-assembly of the PDUs. The MAC layer controls the access signaling across the air interface.

4.6 Security

As we learned about the concept of securing a trusted network for the IMS architecture in Section 3.5, two methods have been acknowledged by the 3GPP to secure access to the

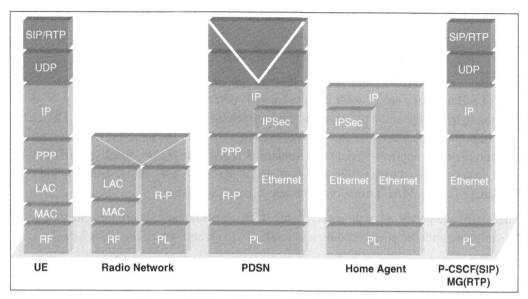

Figure 4.22 Signaling over the CDMA stacks.

network. The first applies to securing at the transport level, the other applies to securing at the protocol level. From an IMS perspective, we need to be able to secure both signaling and media flows. The method should allow a mechanism for key management. Traversing the NA(P)T Section 3.6 should not be an issue. Finally, it should not prevent legal lawful intercept.

4.6.1 TLS

The Transport Layer Security (TLS) protocol uses a reliable transport protocol such as TCP to execute a certificate-based handshake for exchanging encryption keys. It can also be viewed as a method that encrypts a TCP socket. TLS provides three services:

- The mechanism to establish a shared secret key, which is used to encrypt the application data carried in the payload.

- Ensure message integrity by using the shared secret key in the message authentication code (MAC).

- The capability for authentication. It is an asymmetric client-server protocol in which a server first authenticates itself to a client using a server certificate, while it is not essential for a client to present a certificate.

TLS builds the trust relationship with a layered approach with three sub-protocols. The TLS Handshake protocol for key exchange is performed by using public key cryptography for the authentication and cipher exchange to establish a channel. The TLS Record protocol then uses the shared keys exchanged during the handshake to encrypt and compress the application data and transmit it in a secure record. Finally, the TLS Alert protocol is used to notify any TLS-related alerts to the peer entity. These include handshake failure, MAC errors, or certificate-related problems.

The handshake protocol, as shown in Figure 4.23, works on an unencrypted channel. It is initiated with the client sending a ClientHello to create the master secret key. This

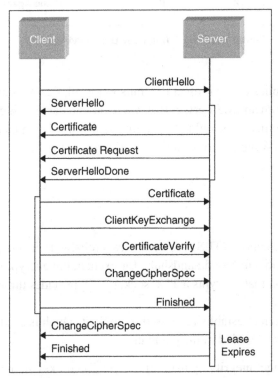

Figure 4.23 TLS—The Handshake protocol.

includes suggested cipher suites and a random number. The server then responds with a chosen cipher suite, server certificate, and a random number. The client authenticates the server certificate sent and proceeds to create the master secret key. The client uses the public key in the server's certificate to encrypt the pre-master secret key generated by the client. The client then exchanges the pre-master key with the server. The server then decrypts the pre-master secret key and generates the master key. This master key is then used in the subsequent encryption.

The TLS record protocol comprises of the steps outlined in Figure 4.24. The application data is fragmented into blocks of 16 Kb size or less. These blocks may be compressed. The MAC is computed for this compressed data using the shared secret key. This fragment is then encrypted with the encryption algorithm using the master key. A TLS header is generated for the fragment describing the content, version, and length. Finally, all the fragments are assembled and transmitted for delivery.

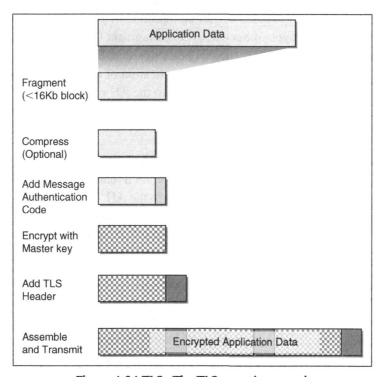

Figure 4.24 TLS—The TLS record protocol.

TLS works with a secure transport protocol such as Transmission Control Protocol (TCP) or Stream Control Transmission Protocol (SCTP). It is suitable for signaling, such as SIP and Diameter, but is not a suitable choice for securing media streams over UDP.

4.6.2 IPSec

IP Security (IPSec) provides a method of encrypting IP packets at the network layer, unlike TLS, which achieves encryption at the transport layer. This makes IPSec more flexible than TLS to handle TCP, SCTP, and UDP transports. IPSec, however, has to rely on an additional protocol for the key exchange. 3GPP has recommended the Internet Key Exchange (IKE) version 2, because of its capability to support the NAT traversal. So, we will be using the term IPSec/IKEv2 when we refer to IPSec or IKE.

IPSec uses the Authentication Header (AH) protocol as specified in RFC 2402 to protect both the IP header and IP data by generating a keyed MAC for most header fields in the IP packet. The fields that are exposed to getting modified by routers are exempt from this. The exempt fields are Type of Service (TOS), Fragment Offset, Flags, Time to Live (TTL), and the IP header checksum. The AH header uses an IPv6 header structure and the same can be used over IPv4. To denote the application of the AH header, the protocol field in the original IP header is set to 51 pointing to the AH protocol. The AH header then carries the original protocol number corresponding to Internet Control Message Protocol (ICMP), TCP, SCTP, or UDP.

For the encryption part, IPSec uses the Encapsulated Security Payload (ESP) as specified in RFC 2406. ESP encrypts only the data portion of the IP datagram with an encryption algorithm. The nature of the ESP header is similar to the AH. The original IP header is set to 50 pointing to the ESP protocol. In addition, an ESP trailer is used to align the data to a 4-byte boundary. Both ESP and AH can be used for a combination of encryption and authentication. ESP is used prior to AH.

IPSec also provides a tunneling option that is widely used. An IP endpoint within a subnet sends its packets to a router performing the function of a Security Gateway (SEG). The SEG takes the complete packet and applies ESP and AH to it. The original IP header is subject to ESP and any local addresses are hidden by the encryption and consequently protected from exposure outside the tunnel. The AH and ESP protocols are shown in Figure 4.25.

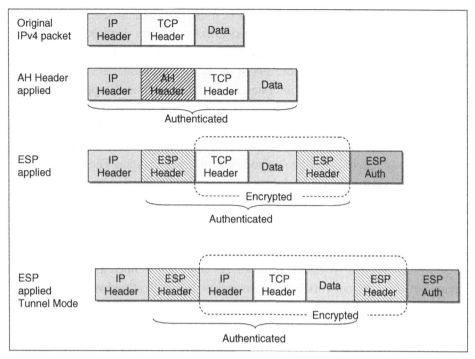

Figure 4.25 IPSec application of AH and ESP.

4.6.3 DIGEST

While TLS finds applicability across several IMS entities for SIP and Diameter, and IPsec provides value for the access network, the scope of Digest is limited to authentication during the registration. Digest is an HTTP security protocol that has been applied to use with SIP. Digest supports the use of a Message Digest 5 (MD5) hashing function on the username/password combination. With Digest extended to SIP, when a SIP client tries to register or establish connection with a UA server, the server sends a 401 unauthorized response to challenge the identity. The client then needs to respond with the credentials.

4.7 QoS

IP networks have evolved with a best-effort data delivery paradigm. This is in contrast with the traditional telecom networks, which have provided guaranteed delivery for signaling and data. There are advantages to a best-effort model, as it keeps the network simple and allows it to scale. While this works effectively with applications not requiring real-time requirements, its weaknesses show up in demanding multimedia applications. The two prominent shortfalls of the best-effort model are delivery delays and packet loss. These manifest as jitter and delays in a multimedia application, resulting in a lower perceived quality to the end user. These problems can be mitigated to an extent by sizing the network and the entities with the right bandwidth. However, the heterogeneity of devices and traffic patterns require a better method to manage the bandwidth available in a predictable and deterministic manner. The QoS protocols augment the best-effort model with enhanced methods of control to provide a guaranteed level of service delivery.

It is intuitive that different applications or network elements have different QoS requirements. There are two basic methods to deliver QoS. These methods can be applied to a single IP flow or an aggregate of IP flows. An IP flow is defined as a unidirectional data stream between two network entities; each flow being identified with a protocol, source, and destination port and address.

- **Resource Reservation** The resources required by an application's request are allocated within the constraints of the bandwidth available. An example is the Reservation Protocol (RSVP).

- **Differentiated Services** The network traffic and resources are allocated by prioritization with demanding applications receiving higher preference. This is applied to an aggregate of IP flows. Diffserv is an example protocol.

The IMS architecture focuses on the guaranteed QoS by applying Policy between the application-specific media requests (SIP/SDP) and the media flows (RTP). Since the handling of the IP flows is done in the IP-CAN, the access gateways to the IP Connectivity Access Network (IP-CAN) such as the GGSN/PDSN perform the policy enforcement. These access gateways may apply RSVP or Diffserv protocols. This is, however, transparent to the IMS Core Network.

3GPP and Telecoms & Internet Converged Services & Protocols for Advanced Networks (TISPAN) also define a set of four QoS classes for transport networks corresponding to IMS services. These requirements must be able to correspond between the IMS network and the IP-CAN for effective service delivery. These four classes are:

QoS Class	Characteristic	Example
Conversational	Delay sensitive but Delay variation sensitive, limited tolerance to packet loss	Audio/Video conversation
Streaming	Delay tolerant but Delay variation sensitive, limited tolerance to packet loss	Streaming Video
Interactive	Round Trip Delay time sensitive, packet content transferred transparently with low bit error rate	Collaboration, Whiteboard
Background	Delivery time insensitive, packet content transferred transparently with low bit error rate	E-mail, Messaging, Chat

4.8 Application

While IMS holds the potential to enable a diverse set of applications for next-generation services, the architecture limits the use of SIP and XML Capabilities Application Protocol (XCAP) to interface with both IMS and legacy application servers. It then becomes the responsibility of applications to determine the protocol interworking to enable access to third-party applications and services. The range of these protocols is vast. Web services are enabled with Simple Object Access Protocol (SOAP), Web Services Definition Language (WSDL), HTTP, and a set of vendor proprietary standards from Google, Yahoo, Microsoft for messaging, and voice interfaces. Legacy protocols need interworking with CAMEL, WIN, AIN, INAP, MAP, and Parlay.

4.8.1 XCAP

The eXtensible markup language (XML) is widely used as a document standard and to represent structured data. IMS utilizes XML for two purposes. The HSS can provide access to the subscriber's profile and service-related information in XML, to a requesting application server or session controller. In traditional cellular systems, this profile would normally be defined in the message protocol structure. This XML data is provided within a text field of a diameter message. The second area of XML usage is to be able to access application-specific data. These are Open Mobile Alliance (OMA)-specified interfaces using the XCAP.

XML, by itself is not a protocol, it is only a data representation scheme. The definition of an XML document or structure is constrained by a Data Type Definition (DTD) or a schema to describe the data. XML is expressed as a set of elements and their attributes.

It maintains a hierarchical organization of the elements expressed as a tree. Tags are identifiers for the elements and their attributes. These are extensible and are defined by the author of the data. The following is a sample of the XML format.

```
<?xml version="1.0" encoding="UTF-8"?>
  <contactInfo>
  <!—Comment: Description --
    <entry>
      <name>Charlie Brown</name>
      <email>cbrown@ims-world.com</email>
        <sip-uri>sips:cbrown@ims-world.com</sip-addr>
        <tel-uri>tel:+1786-951-1242@ims-world.com</tel-uri>
    </entry>
  </contactInfo>
```

The first part of an XML definition, its header, defines the version, encoding, and reference to a DTD or schema. In this data description, we are defining an element for contact information. This is hierarchically organized into sub-elements. The contactInfo has an entry that is described with the name, e-mail, and URI attributes. The values are expressed in free text form. Each tag that represents an element or an attribute has a matching close tag with a "/." XML also supports a mechanism for namespaces. Each tag would be appended with the namespace qualifier. <contact:name> would be used instead in the contact namespace. This allows supporting the same element names from two different sources. For instance, in the preceding example, to include an entry from the billing namespace, we would add the descriptor xmlns:billing=file://billing.xml in the header and use a name entry as <billing:name>.

The second area of use in IMS is to store application configuration data, and access to this data is via XML Configuration Access Protocol (XCAP) defined in RFC 4825. This protocol defines the mechanism to obtain, set, or modify application configuration data from a server where the XML data resides.

4.9 Chapter Summary

IP-based protocols form the basis of communication between the various functional elements in the IMS architecture. The protocols support signaling, media, and a set of specialized functions that includes AAA, security, addressing, and application support. Most session protocols use TCP for the transport. UDP finds its applicability in carrying multimedia and bearer traffic. SCTP is used for securing AAA transports. SIP and Diameter are ASCII-based protocols that are supported by most elements. The SIP/SDP protocol is used for end-to-end session signaling. SigComp provides compression for SIP to improve efficiency while transmitting over a wireless interface. Diameter is used in the IMS Core Network for the AAA function. RTP/RTCP is used for the media plane, with Megaco/H.248 providing the control protocol for the media control functions.

The specialized set of protocols provides the support for addressing, which includes DNS and DHCP. TLS and IPSec are two methods used to secure signaling and media flows. XML is used as a format for sharing complex structured data such as user profiles and service capabilities. The XCAP protocol supports the capability to access and modify this data.

CHAPTER 5

Principles of Operation

We are now getting ready to dive deeper into understanding the principles in the IP Multimedia Subsystem (IMS) and apply some of the concepts we have acquired so far. In this chapter we begin with understanding what is needed for a user to establish one's relationship with the IMS network, to obtain multimedia services. We then probe how the IMS network establishes and controls the user's multimedia session. We examine how these multimedia services are invoked and charged for. And finally, we touch upon aspects of the media processing and security.

5.1 Subscriber Identities and Addressing

In plain voice telephony, a subscriber has been accustomed to having a single public identity. This is typically the Mobile Directory Number (MDN) or the Mobile Subscriber ISDN Directory Number (MSISDN), as used in America National Standards Institute (ANSI) and International Telecommunication Union (ITU) networks, respectively. Most of us are familiar with the 10-digit form of these numbers. The MSISDN, however, is a 15-digit number including the country code also referred to as the E.164 address. The majority of cellular users are oblivious of the fact that they also have a unique private identity for their device. This is the International Mobile Subscriber Identity (IMSI) or the Mobile Subscriber Identity (MSID), as used in ITU and ANSI networks, respectively. This identity is so private that most users have to open the cover of their device to locate it or enter a specific code to find it. Nonetheless, this has great significance to the Mobile Network Operator (MNO) to provision the subscriber in the network.

The concept of a single public and private identity began to change with the advent of e-mail. An identity then assumed the form of a routable address. Aliasing and other means of mail access also introduced the capability to have multiple public and private identities. Extending this further with the advent of Voice over Internet Protocol (VoIP) networks, Session Initiation Protocol (SIP) introduced the concept of a URI, which could be mapped to both a directory number and a routable address. We will see how IMS leverages these fundamental concepts and augments them according to additional usage patterns. Let's get a flavor of some use cases of what the new paradigms of communication have in store so we can accommodate them.

- A subscriber would like to use a single public identity to be associated with more than one device. The subscriber can choose to use one's phone, laptop, Personal Digital Assistant (PDA), or even a TV as a User Equipment (UE) terminal device, all representing the same identity.

- How does a Web-enabled communication service establish a session with an IMS subscriber? How does a skype or a google-talk user call an IMS UE? This requires a service identity in addition to the user identity. Further, this also requires the capability to provide a wildcard to support any authorized user of that service.

- Given concerns for privacy, the identities must be able to support anonymity as well.

A user of the IMS network subscribes to a set of services from the MNO. The identity of the MNO is assigned as the home domain for the user. This is typically of the form *<operatorname>.com*. The service provider must be able to provision the identity and service profile of the user in its Home Subscriber Server (HSS). A public service may also subscribe to the IMS network through a well-defined application server interface. Similarly, the service provider must be able to provision this as well.

We therefore see in Figure 5.1 that an identity in the IMS network can be of two types: one that can be provisioned and the other that is usage-based. Provisioned identity can either correspond to a user or a service. A user must have at least one private identity corresponding to the device and at least one public identity. The service provider normally will assign both a SIP URI and Tel URI for the IMS Public User Identity (IMPU). The IMS Private User Identity (IMPI) is setup as a Network Address Identifier (NAI), in the form *<username>@<operatorname>.com*. Similar to user identity, an external service is identified with a Public Service Identity (PSI). Private service identities have not yet found a suitable case for provisioning at the time of writing.

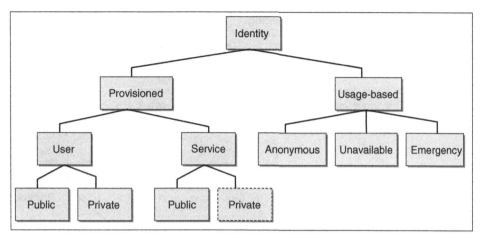

Figure 5.1 Identities in IMS.

There are three identities, also defined for their occurrence as a part of a usage scenario. These have been defined from Release 7 onward. These are not provisioned and occur as a part of an operation. These are *Anonymous*, to support privacy by a user blocking one's identity (e.g., Caller ID blocked); *Unavailable*, for an unavailable identity during a session (e.g., Caller ID unavailable); and *Emergency*, temporarily assigned identity for an emergency call by a non-network subscriber.

To understand the application of these identities, let's consider a use case as shown in Figure 5.2. Mike and Sylvia Green subscribe to IMS services from a (fictional) operator Globus. Mike has a public user identity mike.green@globus.com where other subscribers can reach him. This identity is associated with a private user identity corresponding to the NAI mike@globus.com allocated for his device Id 960119293469 of his imsPhone. Similarly, Sylvia has her public identity Sylvia@globus.com corresponding to her private identity sylvia@globus.com of her imsPhone. The Greens as a couple can be reached at greens@globus.com. This can correspond to either device. Mike subscribes to location services and stock quotes, therefore he has a different service profile.

We now look at an example of a public service profile. For a skype user Bill, to make a call to Mike, the service provider Globus must provision skype (or an equivalent application server) as a public service profile. An identity with a wildcard is allowed. Hence a session invitation from Bill will be seen from user33@skype.com.

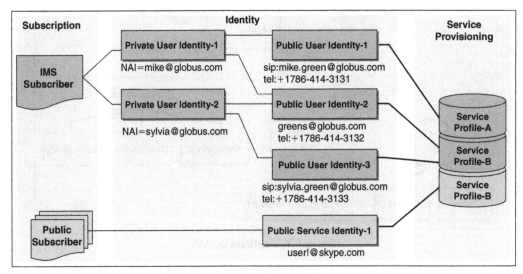

Figure 5.2 User and service identities.

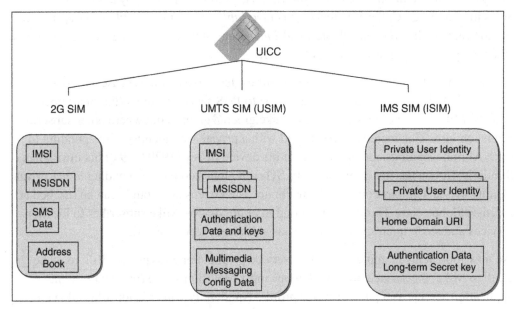

Figure 5.3 The SIM modules.

The usage-based identities take the following allowable forms.

- An identity requesting to be blocked or kept anonymous is expressed as anonymous@anonymous.invalid.

- An unavailable identity is expressed as unavailable@unknown.invalid.

- An identity not subscribed for services on this network requesting an emergency call is expressed as user@sos.domain.

In the situation where the user identity cannot be established, a temporary public user identity based on the IMSI can be substituted.

GSM introduced a powerful concept of a device independent identity. Prior to it, the subscriber identity was tied to a physical device. With the use of a smart card based on a Universal Integrated Circuit Card (UICC), the subscriber identity module (SIM) provides a removable module with the private identity of the subscriber and support for some applications, which have been utilized for SMS data. The SIM module (Figure 5.3) was extended for Universal Mobile Telecommunication System (UMTS) support by Third Generation Partnership Project (3GPP) and enhanced this to the Universal Subscriber Identify Module (USIM). The USIM supports authentication-related information as well. 3GPP augmented this further with the IMS Subscriber Identity Module (ISIM) module with support for public and private user identities.

The ISIM stores the private user identity, which is in the NAI form. It stores all the public user identities for the subscriber as well, which are in the SIP and Tel URI form. It hosts the home domain URI, which is of the form <*operatorname.com*>. Finally, it holds the long-term shared secret key for authentication.

Without the IMS (ISIM) module on the UICC, the User Equipment (UE) can be used for making emergency calls only.

The choice of carrying multiple devices and maintaining a single public user identity posed a problem that was addressed in Release 7 and adopted from draft-ietf-sip-gruu. Consider a situation where a user has, for example, a mobile phone, a laptop, a PDA, a residential wireline phone, and voicemail. Wishing to use the same public user identity for these devices, how should the network determine which device to route the session? To prevent forking of a message to these multiple endpoints, the Globally Routable User Agent URI (GRUU) was introduced.

The GRUU is an identity that identifies a unique combination of Public User Identity and an instance of the UE. This allows a UE to address a SIP request to a specific combination of the Public User Identity and a registered UE. This prevents the request from getting forked to another UE registered for the IMPU. If the UICC is changed, the UE instance is still considered to be the same, so the UE instance ID is dependent on the device and not the UICC.

GRUUs can either be permanent (Pub-GRUUs) or temporary (T-GRUUs). A permanent Pub-GRUU exposes the public user identity and remains in duration across multiple sessions. A T-GRUU is issued for the duration of a registration and does not expose the public user identity. The IMS network can generate the GRUUs. The P-GRUU will always be the same combination for a particular UE instance. The T-GRUU combination will change with each registration. An IMS entity will also be able to derive the IMPU from the P-GRUU, when the P-GRUU is specified in the routing information. The GRUU add the "gr" parameter to qualify the instance of the UA. As shown here, the PUB-GRUU extends the address of record for eric@imsworld.com with an instance number.

```
;pub-gruu="sip:eric@imsworld.com;gr=urn:uuid:
f81d4fae-7dec-11d0-a76500a0c91e6bf6"
;tempgruu="sip:tgruu.7hs==jd7vnzga5w7fajsc7ajd6fabz0f8g5@imsworld.
com;gr"
```

5.2 Subscription in IMS

Having examined the identification for an IMS subscriber, we follow on with how the subscription is maintained in IMS. With reference to Figure 5.4, subscription information is stored in the form of a service profile in the HSS. Each subscriber can have more than one service profile based on the services subscribed. The service profile sets the basis for performing the authorization and authentication functions in the IMS network. The data in the service profile can be accessed by the CSCF functions and the application servers through a diameter-based request. The service profile is exchanged between these elements in eXtensible Markup Language (XML) format.

The subscriber profile provisions both the private and public user identity of the subscriber. Each public user identity is associated with one service profile. However, each service profile can be associated with one or more public user identities. The IMPI is expressed in the usual form of the NAI. The IMPU is expressed in the usual form of the SIP URI and

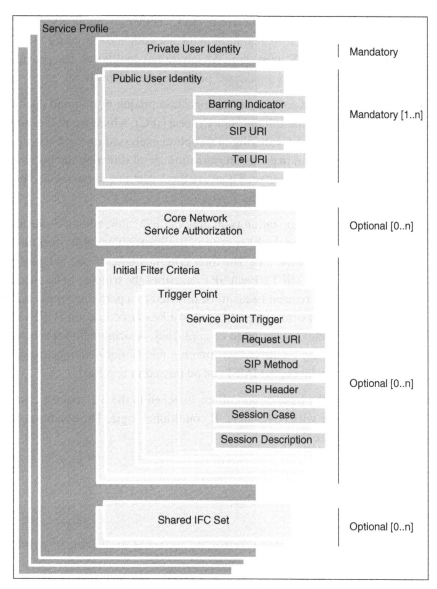

Figure 5.4 Subscription in IMS.

the Tel URI. In addition, the Barring Indicator defines whether the public user identity can be used only for registration and re-registration in the network and barred for other use, or its use is not restricted in the IMS subsystem. An additional qualifier is also provided to identify if the public user identity is a wildcard public service identity.

In addition, the profile contains simple service logic comprising of user and operator preferences stored as a part of the Initial Filter Criteria (iFC), which we will examine in more detail later in this chapter. Filtering is an option exercised on the Serving CSCF, whereby the S-CSCF can make a determination about directing the behavior of the SIP requests to an application server. There are three sections in the service profile corresponding to filter criteria processing.

The Core Network Service Authorization is an optional indicator, which, if present, signals the S-CSCF for exercising the filter criteria for a particular subscribed media. The next section corresponds to the iFC. The filter criteria are described in terms of one or more Service Point Triggers (SPT). Each SPT describes the triggers to be checked to determine whether the SIP request needs to be directed to a particular application server. If the trigger condition is met, the SIP Uniform Resource Locator (URL) of the application server indicates the server to be contacted. A string field for service information is provided for any additional information that is required to be provided to an application server. For instance, the IMSI can be passed in this field.

Triggering can be done on the basis of conditions matched in the SIP request. The conditions can be expressed with AND and OR conditional logic. The conditions that can be matched for triggering are the

- Request URI
- SIP Method
- SIP Header
- Session Case, which describes whether the session is originating or terminating and whether it is registered or unregistered.
- Session Description, which allows you to evaluate each line of Session Description Protocol (SDP) within the SIP message.

Finally, a Shared iFC Set points to a set of initial filter criteria locally administered and stored at the S-CSCF. Shared iFC Sets may be shared by several service profiles.

5.3 Authentication and Authorization

Most IP access networks, including CDMA IP networks, have relied on a single functional element to provide an Authentication, Authorization, and Accounting (AAA) function to terminal device. A single AAA server with a Remote Authentication Dial-in Service (RADIUS) interface has been used to provide the capability for AAA. Authentication refers to the process of providing proof for an identity or source of information and verifying that the identity or information is genuine. The result is establishing a trust relationship with a recognized entity. Authorization, on the other hand, relates to the process of establishing the privileges that can be provided to a recognized entity. Accounting comprises of the functions collecting information on the resource consumption in part of or entire network, related to the service provided to the user.

The AAA function in IMS is shared across two entities. The HSS provides the support for authorization and authentication, and the charging servers provide the accounting function. This enables the capability to provide a better framework to support different accounting methods. We will focus on the authentication and authorization part in this section and discuss charging separately.

5.3.1 The AA Architecture

As we observed in the previous section, the service profile stored in the HSS contains the relevant information about the user identity and the authorized services for multimedia. The authorization task can be enabled by accessing this service-related information. The authentication function requires the additional exchange of the vectors or the triplets to support the challenge handshake. The Diameter protocol enables the session control entities and the application servers to carry out the authorization and authentication function. The I-CSCF and the S-CSCF communicate with the HSS over the Cx/Dx interface, and likewise the application servers communicate over the Sh/Dh interface to obtain the service profile and authorization information. This is shown in Figure 5.5.

5.3.2 The Cx/Dx Interface

The Cx/Dx interface is implemented as a vendor-specific diameter application. Both Cx and Dx share the same command set and Attribute Value Pairs (AVPs). The Dx interface is used when the architecture requires the Subscriber Locator Function (SLF) functional element. In this case, the SLF acts as a Diameter redirect agent, which can

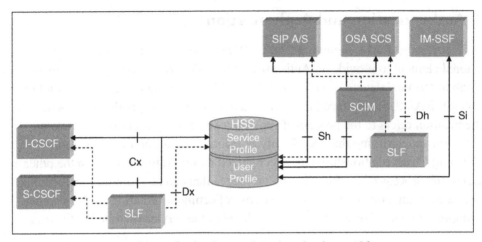

Figure 5.5 The Authorization and Authentication architecture.

route the message to the appropriate HSS, based on the identity of the UE. This interface enables the

- I-CSCF to locate the Serving CSCF for the incoming request

- S-CSCF to obtain the authentication vectors from the HSS

- S-CSCF to obtain the service profile data from the HSS

- S-CSCF to provide the updated registration information to the HSS

- HSS to update the service profile data to the S-CSCF

Both Cx and Dx share a common set of procedures that can be executed between the CSCF (interrogating and serving) and the HSS. The command set in Table 5.1 is used for the registration, location, and service profile access procedures on these interfaces.

The User-Authorization, Server-Assignment, and Multimedia-Authentication are used in the registration process. The Registration-Termination is used for de-registration. The Location-Information is used to determine the URI for the S-CSCF or Application Server (AS) for direct routing. The Push-Profile is used for downloading the service profile to the S-CSCF, when it is updated.

The I-CSCF sends a UAR to the HSS to authorize the registration of the public user identity, access permissions, and roaming agreements. It also obtains the address of the

Table 5.1 Diameter Cx/Dx interface commands.

Command	Description	Direction
UAR	User-Authorization-Request	I-CSCF → HSS
UAA	User-Authorization-Answer	HSS → I-CSCF
SAR	Server-Assignment-Request	S-CSCF → HSS
SAA	Server-Assignment-Answer	HSS → S-CSCF
LIR	Location-Information-Request	I-CSCF → HSS
LIA	Location-Information-Answer	HSS → I-CSCF
MAR	Multimedia-Authentication-Request	S-CSCF → HSS
MAA	Multimedia-Authentication-Answer	HSS → S-CSCF
RTR	Registration-Termination-Request	HSS → S-CSCF
RTA	Registration-Termination-Answer	S-CSCF → HSS
PPR	Push-Profile-Request	HSS → S-CSCF
PPA	Push-Profile-Answer	S-CSCF → HSS

S-CSCF, which will serve this request. The HSS also performs an initial check if the public user identity corresponds to the private user identity sent in the message.

The S-CSCF sends the MAR to HSS to obtain the authentication vectors from the HSS, and to resolve any synchronization issues between the sequence numbers in the HSS and the UE.

The S-CSCF sends the Server Assignment Request (SAR) to HSS to register the S-CSCF to a public user identity, or deregister the S-CSCF assigned to one or more public user identities. It also enables the S-CSCF to obtain the service profile and charging information regarding the public user identity.

The I-CSCF sends the LIR to the HSS for the purpose of direct routing. The I-CSCF uses the command to obtain the address of the S-CSCF, where the public user identity is registered from the HSS.

To clear the registration of public user identities for administrative purposes, such as for bulk de-registration, the HSS sends the RTR to the S-CSCF. This notification indicates to the S-CSCF to de-register the public user identities.

The PPR is also initiated by the HSS. When the service profile or charging information is updated for a user, it will download the updated service profile and/or charging information to the S-CSCF. Since the service profile is identified by the public user identity, the Push-Profile request uses the private user identity, in contrast to the public user identity used for most of the messages discussed previously.

The following set of AVPs as shown in Table 5.2 used in the procedures just discussed.

Table 5.2 Diameter Cx/Dx interface AVPs.

AVP	AVP Code	Purpose
Visited-Network-Identifier	600	Contains an identifier such as a domain name, to identify the visited network
Public-Identity	601	The public user identity in the SIP URI or Tel URI format
Server-Name	602	The SIP URL to identify a SIP Server, such as a S-CSCF
Server-Capabilities	603	A Grouped AVP that assists the I-CSCF in the S-CSCF selection
Mandatory-Capability	604	An operator assigned identifier for determining the S-CSCF capability
Optional-Capability	605	An operator assigned identifier for determining the S-CSCF capability
User-Data	606	This AVP contains the information that the S-CSCF needs to give service to the user
SIP-Number-Auth-Items	607	Indicates the number of authentication vectors the S-CSCF is requesting
SIP-Authentication-Scheme	608	The scheme used in the authentication of the SIP messages
SIP-Authenticate	609	Contains specific parts of the authenticate data portion SIP headers that are expected in the SIP response
SIP-Authorization	610	Contains specific parts of the authorization data portion SIP headers that are expected in the SIP response
SIP-Authentication-Context	611	Contains Authentication-related information that is not a part of the SIP authentication headers
SIP-Auth-Data-Item	612	A grouped AVP containing the authorization and authentication information for the requesting client
SIP-Item-Number	613	Used in the SIP-Auth-Data-Item, provides the sequencing when multiple SIP-Auth-Data-items are used

Table 5.2 (continued)

AVP	AVP Code	Purpose
Server-Assignment-Type	614	Indicates the type of server update being performed in a SAR operation, such as the registration, de-registration, and so forth
Deregistration-Reason	615	Indicates the reason for de-registration
Reason-Code	616	Provides the reason for the network initiated de-registration
Reason-Info	617	Text information to notify the user about de-registration
Charging-Information	618	A grouped AVP containing the address of the charging functional elements
Primary-Event-Charging-Function-Name	619	Contains the URI of the primary online charging function
Secondary-Event-Charging-Function-Name	620	Contains the URI of the secondary online charging function
Primary-Charging-Collection-Function-Name	621	Contains the URI of the primary charging data function
Secondary-Charging-Collection-Function-Name	622	Contains the URI of the secondary charging data function
User-Authorization-Type	623	Indicates the type of authorization being requested in a UAR operation
User-Data-Already-Available	624	Indicates to the HSS, if the S-CSCF already has a part of the service profile already available
Confidentiality-Key	625	Contains the Confidentiality Key (CK) for authentication
Integrity-Key	626	Contains the Integrity Key (IK) for authentication
Supported-Features	628	This grouped AVP identifies to the destination about the features the request originator supports
Feature-List-ID	629	Part of the Supported-Features, identifies a feature list
Feature-List	630	Part of the Supported-Features, describes the feature list
Supported-Applications	631	This grouped AVP contains the supported application identifiers of a diameter node for the accounting and authentication apps and any vendor specific
Associated-Identities	632	This grouped AVP contains private user identities associated with an IMS subscription

5.3.3 The Sh/Dh Interface

The procedures on the Sh/Dh interface enable the AS to perform authorization and authentication. The AS must be able to authorize UE requests by validating against their subscribed services. Since the AS receives requests from a public service identity for a service from the IMS network, it must be able to enable authentication for this identity. Additionally, the procedures allow the AS to support updates to service data, which is a powerful feature that can allow user control of subscribed services. This interface thus enables the AS to receive the service profile and charging data from the HSS and provides updates to the HSS when the service profile is updated. It also enables the AS to receive notifications when any service profile data changes take place in the HSS.

To support the procedures for enabling service level authorization, service level authentication, and user data access, the command set in Table 5.3 is used in the Sh/Dh interface.

Before we explore these Diameter commands, we need to learn a little more about a special data element—the User Data. Similar to the service profile on the Cx interface, this AVP transports the service-related information in an XML format between the HSS and the AS. However, the User Data is a broader set of data than the service profile, and provides more details for describing a user profile.

The logical description of the User Data, as depicted in Figure 5.6, contains the user-specific data such as the user identity, service data, location information, and state of the

Table 5.3 Diameter Sh/Dh interface commands.

Command	Description	Direction
UDR	User-Data-Request	AS → HSS
UDA	User-Data-Answer	HSS → AS
PUR	Profile-Update-Request	AS → HSS
PUA	Profile-Update-Answer	HSS → AS
SNR	Subscribe-Notifications-Request	AS → HSS
SNA	Subscribe-Notifications-Answer	HSS → AS
PNR	Profile-Notification-Request	HSS → AS
PNA	Profile-Notification-Answer	AS → HSS

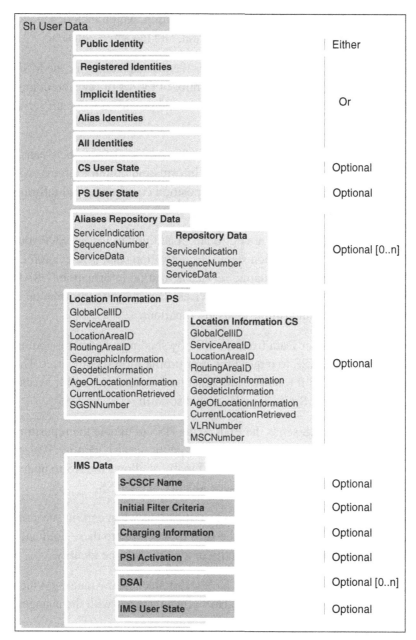

Figure 5.6 The user data on the Sh interface.

user. In addition, it contains the IMS service information, which includes the S-CSCF address, initial filter criteria, charging information, and the activation state.

The request for the User Data typically contains the private identity or the MSISDN. The User Data returns the corresponding public identity or the set of registered, implicit, and alias identities. For a PSI request, the User data would provide a wildcard service identity if configured at the HSS.

The Repository data contains transparent data that can be exchanged between different AS supporting the same service. The Location Information both for circuit-switched and packet-switched network support provides the position coordinates and information about the serving network entities.

In the IMS data, the returned fields are optional. In addition to the S-CSCF address and the iFC, this set contains an activation state. The activation state is relevant for both a public service identity and the dynamic service activation information (DSAI). It corresponds to the state for the subscribed services. The charging information contains the addresses of the online and offline charging functions.

The AS can request this User Data from the HSS by sending a Diameter command. The AS sends the UDR to the HSS to request the user profile for a specific user. The AS will indicate the identity set it requires, and other optional information it needs such as location, service activation, S-CSCF address, and so forth.

For other procedures, the AS sends the PUR to the HSS to update the repository data stored in the HSS for each public user identity or public service identity. It also permits the AS to update the DSAI stored in the HSS. Finally, it allows the AS to update the PSI activation state for a particular public service identity in the HSS.

The AS can register with the HSS to receive notifications when certain subscription data changes in the HSS. The SNR is sent to the HSS to subscribe to these notifications. The data can be selected for a public user identity or a public service identity.

To inform the AS about any changes that have taken place in any data for which the AS registered to receive notifications, the HSS sends the PNR with the changes in the user data.

Table 5.4 describes the AVPs used in the previous commands.

Table 5.4 Diameter Sh/Dh interface AVPs.

AVP	AVP Code	Purpose
User-Identity	700	This is a grouped AVP, which contains either the public user identity in the form of the SIP or Tel URI, or the MSISDN of the UE.
MSISDN	701	The MSISDN is useful as an identifier for applications that inter-work with the current generation of mobile devices.
User-Data	702	This AVP contains the user profile data that is used in the UDR, SNR, and PUR operations. This contains the XML format of the data described earlier.
Data-Reference	703	Indicates the type of User data requested in the UDR and SNR operations.
Service-Indication	704	This AVP identifies a service in the AS.
Subs-Req-Type	705	This indicates for a subscription request, whether to subscribe or unsubscribe to notifications.
Requested-Domain	706	This qualifies the data is being requested for the CS or the PS domain.
Current-Location	707	Indicates whether to initiate an active location retrieval.
Identity-Set	708	Defines the set of identities—all, registered, implicit, or alias.
Expiry-Time	709	Defines the expiry time for subscriber notifications.
Send-Data-Indication	710	Indicates to the HSS whether to send the User profile data.
DSAI-Tag	711	Contains the identifier for the dynamic activation instance.
Public-Identity	601	As per the Cx interface.
Server-Name	602	As per the Cx interface.
Supported-Features	628	As per the Cx interface.
Feature-List-ID	629	As per the Cx interface.
Feature-List	630	As per the Cx interface.
Supported-Applications	631	As per the Cx interface.

5.4 Session Control

While every component in the IMS architecture plays a vital role in its operation, session control is the core function. Just as the legacy telecom networks have been built on the foundation of switching and call control, the heart of IMS is session control. It is apparent that IMS is about multimedia sessions, so control of the sessions is essential. However, as we will see in this section, session control differentiates itself from other session architectures such as vanilla SIP VoIP and Peer to Peer (P2P). Session control is closely tied in with service control, in its capability to invoke services and facilitate service interaction.

5.4.1 Session Control Architecture

SIP is a fundamental building block of session control in IMS. At first glance at Figure 5.7, adopting a SIP session control model conceptually resembles the SIP trapezoid paradigm we saw earlier in Figure 4.5. However, there are differences as well. Recall that the trapezoid model was about using proxy and locate functions to establish a handshake between two user agents, until they can communicate directly with each other and establish their direct bearer path. In IMS, the P-CSCF plays a similar role

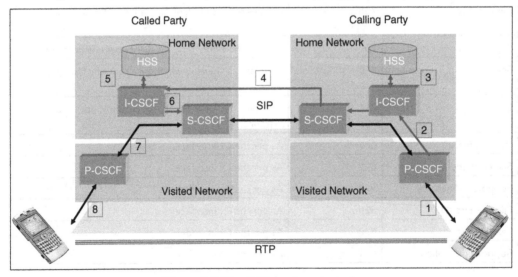

Figure 5.7 The IMS session control.

to SIP proxies to enable the UE-UE signaling through disparate networks. The bearer connection is directly set up between the two UEs. The difference, however, lies in the session control that resides within the IMS core network. In the SIP trapezoid model, the signaling subsequent to the handshake between the two user agents takes place directly. In IMS, the signaling will always take place through the core network. One main reason is being able to modify the session under Quality of Service (QoS) control, unlike the SIP model, which assumes best effort. This is the model of the session control.

Let's follow from Figure 5.7 how the session control gets established for a basic UE-UE communication. Consider a scenario with the UE roaming in a visited network. The UE requests service from the visited network. It initiates an IP-CAN session and establishes initial communication with the P-CSCF. The P-CSCF determines the home network of the UE by its public user identity. The P-CSCF proxies the request for a session to the home network. This request is handled by the I-CSCF, which needs to determine the S-CSCF that will provide the control for this session. The I-CSCF obtains the address of the S-CSCF from the HSS. It communicates the session request to the S-CSCF. The S-CSCF then needs to determine the home network of the called party, where the session request can be directed. The session request is handled by the I-CSCF in the called-party home network, which determines the serving S-CSCF. The S-CSCF then directs the call toward the P-CSCF, where the called party is visiting. Upon acknowledgement from the called party, the session control for establishing the call is exchanged through the S-CSCF. Once the handshake is established, the bearer or the Internet Protocol (IP) flows set up independently on the IP-CAN. The session is now established.

It has become apparent to us by now that actors in the session control stage are the CSCFs. Having had an initial capability view of these elements, let's explore them further.

The P-CSCF plays a SIP proxy role in session control. It is the first point of access into the IMS core network for a UE request. In normal processing, it forwards the requests made on behalf of the UE without modifying the Request-URI. Only under abnormal cases can it originate or terminate a SIP request. The P-CSCF forwards an initial request such as a REGISTER to the I-CSCF, as determined by the UE-provided home domain name. Subsequent UE requests are forwarded to the S-CSCF, following a successful registration. The P-CSCF forwards any network-generated requests or responses to the UE. The P-CSCF also provides session control to detect and handle emergency sessions. As discussed in Chapter 4, it also provides the edge access functions for security, QoS, and compression.

The I-CSCF plays an enabler role in session control. The function of the I-CSCF is primarily to enable the selection of and routing toward a selected serving S-CSCF for a SIP request. The I-CSCF needs to interrogate the HSS to make the determination that S-CSCF should be selected for session control. Upon receiving a registration request from the UE, the I-CSCF determines and assigns the S-CSCF. The I-CSCF also enables the routing to the S-CSCF, for an initial SIP request received from a different network. The I-CSCF also needs to perform a translation function before it interrogates the HSS. It needs to translate the E.164 address in the Request-URI into the Tel URI format.

While the P-CSCF and I-CSCF play a supporting role, the S-CSCF performs the session control function. The S-CSCF has full capability to accept, forward, originate, and terminate SIP messages. It provides both a call model and state model for managing inbound and outbound dialogs. The S-CSCF can provide session control by operating in different personalities of SIP servers. The S-CSCF

- may function as a SIP registrar by accepting the UE registration requests and updating the information in the HSS (which functions as the location server).

- may function as a proxy server by accepting incoming session requests and forwarding them to a destination such as an application server, Border Gateway Control Function (BGCF), or an Media Resource Function Controller (MRFC).

- may function as a User Agent and may terminate and originate SIP transactions.

The S-CSCF handles processes session-related requests from two originating endpoints— the UE and the AS. The behavior for session control on these requests is as follows.

UE Request
- Determine if the request is to be filtered for matching service trigger conditions. The request needs to be routed to an application server.

- For a request not requiring any service related processing, determine the situation of the destination endpoint. If the destination UE belongs to a different network, the request needs to be forwarded to the I-CSCF of the home network of the destination UE.

- If the destination UE belongs to the same network, the request needs to be forwarded to a P-CSCF of the same network.

- If the destination endpoint corresponds to a Tel URI and needs to be routed to the circuit-switched (CS) domain, the session request is forwarded to the BGCF for subsequent routing.

- Apply any required translation for the routing information in the Request URI for E.164 translation.

- If media resources are being requested for the session—announcements or tones—the request is forwarded to the MRFC.

Application Server Request

- AS requests also follow similar processing. The S-CSCF needs to determine the user on whose behalf the AS is originating the request.

- Determine any loop conditions that may get caused by AS to AS requests

5.4.2 Leg State and Control Model

A session control model and service control is constructed to manage two aspects of the session—a unit in the session and the states it goes through. The concept has evolved from traditional telephony call control. This defined a leg as a unit in the call and the various states of the call such as setup, answer, teardown, and so forth. A typical call has two legs: incoming and outgoing. A multiparty call such as a conference may have three legs. Although the 3GPP standards still define the unit of a session as a leg, the SIP RFC has deprecated the former use of a "call leg" and refers to it as a dialog instead. In this section, we continue to use the term "leg," which is the same as a SIP dialog.

The Leg Management applies to both session-related flows and non-related flows such as registration. There are two models for Leg Management as seen in Figure 5.8. The Leg State Model (LSM) is responsible for maintaining state information about a leg and propagating it as necessary. The LSM determines whether state information is to be explicitly changed while passing it to the destination or implicitly routed. The behavior of the LSM is according to the function of a proxy, redirect server, or UA being applied in the S-CSCF for the session. The LSM is constructed as a combined model for the incoming and outgoing legs.

The Leg Control Model (LCM) is responsible for acting on the SIP information exchanged on a leg with the destination endpoint. The Incoming LCM, for instance, provides control on the information exchanged between an incoming SIP request from

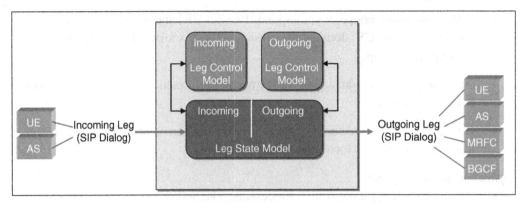

Figure 5.8 The Leg State and Control Model.

a UE to an AS. The outgoing LCM, for instance, provides control on the information exchanged between an incoming SIP request from an AS to a UE.

5.4.3 Forking

Forking a request is a powerful feature of session control in IMS, which applies the concept of SIP forking. This is the capability for a proxy server to fork a message to multiple destination endpoints. The S-CSCF operating in a proxy mode provides this capability to forward an incoming request to multiple destinations. The major benefit forking provides is the capability to direct a request to multiple destinations corresponding to a single public user identity. Consider an instance where a subscriber has registered one's IMS Public Identity (IMPU) with two endpoints, such as one's mobile phone and a laptop. An incoming call to this user's IMPU can be forked by the S-CSCF to both endpoints simultaneously. The user can answer either through a softphone client on the laptop or the cell phone, whichever is convenient.

Forking can be done either sequentially or in parallel. This preference for contact addresses can be set by the user upon registration. If this is not provided, the S-CSCF uses its default for configuration.

5.5 Charging and Billing

We now continue the AAA function and understand how the charging in IMS provides a sophisticated system for accounting. Charging provides the functions to collect data on the usage of network resources for service delivery, and the capability to bill the subscriber for the delivery of these services. Third-generation (3G) systems provide a layered charging model, where these functions can be implemented across three network levels. Charging can be done on the bearer, network (subsystem), and service levels. Charging at the bearer level, such as at the Gateway GPRS Support Node (GGSN)/Packet Data Serving Node (PDSN), enables methods such as flow-based charging. Charging at the network or subsystem level enables both online and offline methods for generating event- and session-based information. Charging at the service layer, such as messaging, the location provides a more granular unit of service delivery to the user.

While the charging functions are implemented within the IMS network elements, the billing function is handled within a separate billing domain. This function contains the necessary rating functions as applied to the charging data received, account balance management, and all the back-office processing to generate billing information for the subscriber.

Offline charging is the process by which resource usage information is recorded along with the service execution, and stored for delivery to the billing functions. In addition, online charging has the capability to change the service execution in real-time, based on obtaining authorization for the resource usage. Offline charging therefore focuses on generating event records and the transfer functions to the billing system. Online charging, on the other hand, provides the mechanism to bind volume of data or duration to a charging request and authorize the requisite resource usage for service execution. Offline charging has typically been applied to postpaid billing, and online charging has been typically utilized for credit based or prepaid billing.

Online charging is applied for both session-based charging and event-based charging. Session-based charging applies to charging for multimedia sessions, Packet Data Protocol (PDP) contexts, and so forth. Event-based charging is used for content based charging models.

Flow-based charging is useful for determining charging of IP-flows or at the packet level. This function is enabled at the bearer handling layers (i.e., at the GGSN, PDN, or equivalent IP-CAN mechanism).

5.5.1 The Charging Architecture

The Charging architecture in the IMS Core Network supports both offline and online charging methods. The Charging Trigger Function (CTF) is a common function to both methods that generates the charging events based on the usage of network resources. This function is co-resident with the IMS functional elements, and its scope depends on the type of signaling or traffic the element provides. As described in Figure 5.9, the CTF is implemented within the signaling elements of the CSCF, and media functions of the MRFC, MFCF, BGCF, and the SIP AS. The CTF within these elements enables the collection of the relevant accounting metrics that are triggered on a particular event (e.g., starting of a SIP session). Once the information about the event is triggered, it is collected and recorded and made available over the Rf interface to the offline data processing elements, and over the Ro interface to the online charging system.

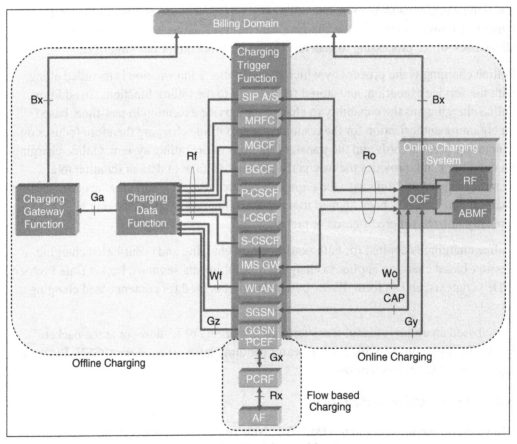

Figure 5.9 The Charging architecture.

In the offline charging method, the Charging Data Function (CDF) obtains the charging events from the CTF and utilizes the same to construct Call Detail Records (CDRs). The CDF may employ different methods to generate the CDR. A CDR may be generated from a single charging event, or it may derive inputs from multiple charging events.

The CDF forwards the CDRs that have been generated immediately to the Charging Gateway Function (CGF) over the Ga interface. The CGF provides any pre-processing, validation, and reformatting of the CDRs before these are packaged and transferred to the billing domain over the Bx interface. The CDF also provides persistent storage for the CDRs and the necessary file management and transfer operations.

In the online charging method, the CTF plays a different role than the offline counterpart. The CTF is more than a recorder function, and can support the capability for service control. To do so, it must be able to obtain authorization for resource usage from the Online Charging Function (OCF). It must be able to put service processing on hold until resource authorization has been received. On receiving the permission, it must be able to direct the resource usage and monitor for any quotas or reservation. Finally, it must be able to terminate resource usage on authorization refusal or expiry.

The Online Charging System (OCS) contains the OCF. The OCF has the capability to support both session-based charging and event-based charging methods. In addition to the OCF, the OCS also contains the Rating Function (RF), which provides a monetary or non-monetary value to the charging event. This rating can apply for data volume, session, or service events. Finally, the OCS also provides the Account Balance Management Function (ABMF) for ensuring a balance check and a balance return for the service requested.

5.5.2 Charging Principles

The generation of charging data for both online and offline charging is done based on two methods: event-based charging and session-based charging. A chargeable event in event-based charging is a single user-to-network transaction. Sending a multimedia message is an example. This chargeable event results in a single CDR generation for offline charging, and a single authorization request for credit control in online billing. In the offline charging model, the charging event generated by the CTF within the network element is forwarded to the CDF on the Rf interface. The CDF then generates a corresponding CDR, which is transferred to the Charging Gateway Function (CGF) on the Ga interface. This CDR is eventually handed over to the billing domain over the Bx interface.

For the online charging model, when the charging event is detected by the CTF, it sends a credit control request to the OCS on the Ro interface. The OCS makes a determination about sufficient credit availability for this charging event. The online charging function within the OCS makes a determination about credit availability with the help of the rating function and account balance management.

Session-based charging applies to a user session such as an IMS SIP session, voice call, or GPRS PDP context. A user session may typically result in more than one chargeable event resulting in the generation of more than one CDR for offline charging and multiple interactions for credit control. In the offline charging model, the initial charging event generated by the CTF is forwarded to the CDF on the Rf interface. During the course of the session, other chargeable events such as an answer or termination are forwarded to the CDF. The CDF binds these events into one or more CDRs based on the type of session and transfers to the CGF for distribution to the billing domain.

In the online charging model, the CTF requests credit authorization from the OCS over the Ro interface. The network element will have to hold the session from proceeding until it gets a successful response from the OCS. To permit the session, the OCS may respond with a reservation of units from the subscriber's account for the session to proceed. The network element can translate these units to resource usage. If this quota is exceeded, the CTF may notify the OCS with an interim event. Once the session is terminated, the CTF reports the final charging event for the OCS to return any unused quota to the subscriber's account. For each charging event that is reported, if the OCS determines that there are insufficient credits, it may direct the session to an announcement or credit recharging function.

We will examine flow-based charging in more detail in the next section.

5.6 Policy and QoS

Service-based policy for QoS control was introduced in the Release 5 specifications. Since then it has evolved both in terms of its scope and definition. The initial policy mechanisms utilized an Internet Engineering Task Force (IETF) Common Open Policy Service-PRovisioning (COPS-PR)-based interface, also referred as the Go interface. The method to apply policy-based control for QoS relied primarily on a policy decision point and a policy enforcement point. This has been extended to an enhanced policy control framework with requirements from different types of IP-CAN. Further, to extend

IP flow-based charging to a greater degree of support for accounting and credit control, the interfaces evolved to using diameter instead of COPS. Diameter can provide complete AAA support functionality as opposed to COPS.

5.6.1 The Policy and Charging Control Architecture

The application of service-based policy to ensure QoS is a three-step process. First, a determination needs to be made on the signaling plane about the QoS being requested for the multimedia session. A QoS request can be in terms of bandwidth required for uplink and downlink for the communication channel. A decision then needs to be made for the allowance of the QoS. This requires the application of local policy rules that are defined by the service provider.

Policy is enforced on the user plane. This is achieved by a gating mechanism applied to the IP-flows. The gate can be applied to an individual or a set of IP-flows. This gate is implemented in the bearer handling IP-CAN function, which is the GGSN/PDSN. A gate comprises of a packet classifier and a gate status. When a gate is open, the packets matching the packet classifier are permitted for any resource reservation. When a gate is closed, the packets matching the packet classifier are dropped.

Guaranteeing QoS requires ensuring the resource availability and bandwidth control for the duration of the multimedia session. The initiator of the session—the UE for a UE originated call and the S-CSCF for a UE-terminated call—needs to make the request. While setting up a SIP session, the initiating UE or the S-CSCF signals in the SDP message about the resources and bandwidth that are being requested for this session. Ensuring this request requires the following. Is the subscriber authorized for the resources requested? Are these available? If so, then how to coordinate with the traffic plane?

As we see in Figure 5.10, the signaling plane is handled by the P-CSCF, also termed as the Application Function (a term that reflects the equivalence with the Telecoms & Internet Converged Services & Protocols for Advanced Networks (TISPAN) architecture). The P-CSCF is restricted to proxy functions in the signaling plane. To supplement, the policy architecture introduces a set of new elements to coordinate this interaction between the signaling and the bearer plane.

The Application Function (AF) provides the translation between Session Description Protocol (SDP) parameters at the application level to policy setup information. Over the

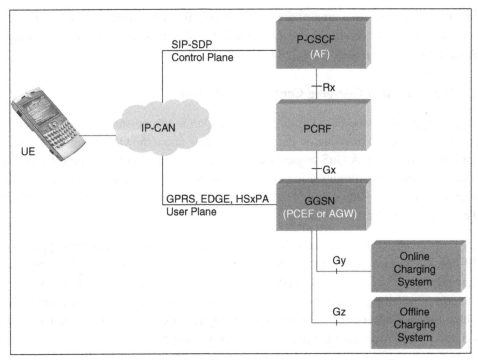

Figure 5.10 Policy architecture in the UMTS IP-CAN.

Diameter Rx interface, it communicates this information to the Policy And Charging Rules Function (PCRF).

The PCRF provides a rules-based function for the gating of IP-flows on the IP-CAN access gateway (AGW). As mentioned earlier, it extends the charging control to the IP-flows. The PCRF is responsible for making the policy control decisions and communicates it to the enforcing function on the AGW on the Diameter Gx interface. The PCRF makes the policy decisions based on the parameters received from the AF, and from operator configuration. The PCRF maps this information to the QoS parameters for the IP-flows.

The Policy Charging and Enforcing Function (PCEF) resides on the gateway to the IP-CAN, which in the UMTS networks is the GGSN. The PCEF enforces the QoS directives it has received from the PCRF. The PCEF can either request policy decisions from the PCRF, or these can be pushed to it. This provides the policy-based admission control that is applied to the user-plane bearer and the gating of IP-flows. This mechanism

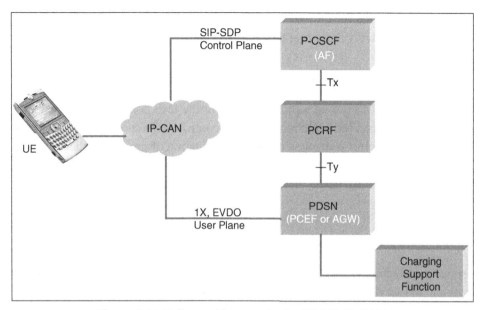

Figure 5.11 Policy architecture in the CDMA IP-CAN.

ensures that the resources that can be used by a set of IP flows are authorized by the policy set via the AF request and determined by the Policy and Charging Rules Function (PCRF).

The resource reservation for use by the PCEF is expressed by the PCRF in terms of authorized bandwidth and a QoS class. The QoS class is an identifier for a bearer service. The PCRF determines the maximum authorized bandwidth for the QoS class. This information is mapped to the admission control function in the IP-CAN element.

Applying the charging control function in association with the policy enforcement allows the service provider to monetize the QoS function. Policy rules can be constructed to suitably charge for the QoS rendered for the multimedia session. The PCEF can communicate with the online charging system over the Gy diameter interface for the credit control application. This is similar to the Ro interface, but applied to data flow-based charging. The Gz diameter interface between the PCEF enables the recording of IP-flow charging events, which are forwarded to an offline charging system for CDR generation.

Figure 5.11 describes the architecture in the CDMA context. 3GPP2 defines the interaction between the P-CSCF(AF) and the PCRF over the Diameter Tx interface.

The PCRF communicates the QoS directives and policy decisions to the PCEF function, which resides in the PDSN providing the IP-CAN function.

5.7 The Service Plane

The Service Plane in IMS provides the flexibility for the IMS core network to extend application logic to affect the behavior of session control. In the process of the session control, the service control is invoked on the determination of pre-established criteria. This control is directed from the S-CSCF to the application servers over the IMS Service Control (ISC) interface. By virtue of this service control, new applications can be effectively created.

5.7.1 The Service Architecture

The service architecture of IMS provides a clear separation between the session control elements and the service logic, as noted in Figure 5.12. This boundary helps to enable new services without requiring upgrades to session control elements. Control for the

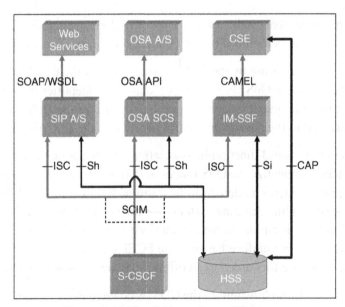

Figure 5.12 The IMS service architecture.

session is transferred to the service layer. The service layer thus has the capability to make decisions on the session behavior.

With a SIP enabled architecture, new service logic for IMS is best hosted on SIP application servers. Accommodating existing service logic, however, is a challenge. The standards authors had been cognizant of the existing revenue-generating services that exist in the current generation of networks. It becomes essential for IMS to co-exist with these services. Thus in addition to SIP AS, the service layer must be able to support the legacy service environment. There are two types of services: the Intelligent Network (IN) services, which are enabled by Customized Applications for Mobile Networks Enhanced Logic (CAMEL) in the International Telecommunications Union (ITU) networks, and by WIN in the ANSI networks. These services include toll-free services, short code dialing, and local number portability VPN, to name a few. In addition to IN, certain service logic is implemented as OSA services.

The service layer, therefore, must allow service gateways or interworking elements to provide a mapping between the SIP domain and IN/OSA domains. The Open Service Access (OSA) service capability server (OSA-SCS) provides the gateway function to the OSA application servers using the OSA/parlay APIs. The IMS service switching function (IM-SSF) provides an emulation of an SSF function to an IN service control point (SCP). The IM-SSF must be able to provide a translation between SIP and IN protocols such as CAMEL/WIN.

All three types of service elements interact with the session control layer using SIP. These elements communicate with the S-CSCF using SIP on the ISC interface.

Since the driving force for IMS has been to enable convergence with the IP world, the SIP AS provides the leverage to interface with services from the IP domain. These services are the Web services enabled with the Simple Object Access Protocol (SOAP) and Web Services Definition Language (WSDL) interfaces. The IMS standards do not specify the interfaces the AS can support, extending to other domains. Nor do the standards specify how the service logic needs to be implemented for the inter-working. This is left to the discretion of the implementers. The important facet of SIP AS is that this flexibility also allows interfacing to the service-oriented architecture (SOA). We will examine this in Section 8.2.

To obtain service authorization, the application servers communicate with the HSS over the Sh interface. As we observed earlier, this is typically to obtain the user profile.

This interface also provides a powerful feature of being able to update the user profile. For SIP application servers enabled with a secure Web services interface, this can result in the subscriber controlling one's service features. An example of this would be time-of-day routing.

Finally, the service architecture also supports the capability to provide a new dimension of service interaction with the Service Capability Interaction Manager (SCIM). We will investigate this further in the next section.

5.7.2 Service Invocation

As we recall from Section 5.2, the invocation of services is performed from the session control layer. This requires prior knowledge of the subscription and a trigger-detection condition that a service is being requested. This takes place in the S-CSCF session control element and is achieved by the combination of the Service Point Triggers (SPTs) and the Initial Filter Criteria as shown in Figure 5.13. The service profile, which is expressed as a set of the iFC, is downloaded to the S-CSCF from the HSS on the Cx interface. This directs the S-CSCF to filter the behavior of session initiation to determine if any service requiring conditions are being triggered.

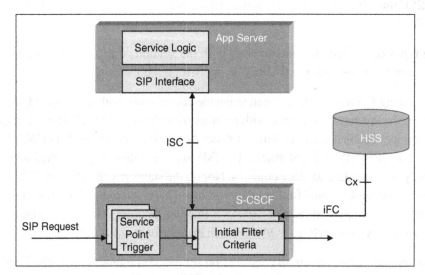

Figure 5.13 Service invocation.

The elements in the filter criteria that play a role in determining the invocation of the service plane are:

- The address of the application server, which will be invoked if the conditions in this filter criterion are matched.

- The default handling to be performed, if the request cannot reach the application server.

- The priority of the iFC, which is essential to provide sequencing when a sequence of iFC has to be traversed.

- The conditions for the triggering, expressed as SPTs, that must be matched for the service invocation.

SPTs can be applied to SIP signaling on which the iFC can be applied. These SPTs are:

- Any initial SIP method; e.g., an INVITE.

- The type of registration in a REGISTER request.

- The SIP Header field. This can be matched to the presence or absence of a header field. Both type and content can be matched.

- The originator of the request; i.e., the Request-URI

- The direction of the request, whether it is UE originating or UE terminating.

- The session description; i.e., an SDP field.

Some general rules of the iFC processing are:

- Filter criteria can be applied in a single number or as a set.

- Filter criteria support sequencing, which implies that multiple filter criteria can be applied to a single request. The traversal of a set of filter criteria is based on the priority assigned to each iFC, from highest to lowest.

- The Default handling in the iFC allows continuation to the next iFC or exiting the set, as defined in the iFC.

- iFC processing for the initial event. REGISTER differs slightly from the other processing to enable sending a third-party REGISTER to the application servers.

- Filter criteria processing should be carefully done to avoid looping. A SIP request from an application server should be refrained from sending back to the same server on an SPT condition being met.

Let's examine the sequence of steps involved in the service invocation process with the iFC. The S-CSCF

1. Receives the filter criteria from the HSS corresponding to a service.

2. Will set up the filter criteria for each request type that needs to be filtered for service-level processing.

3. Interprets an incoming SIP request and applies the highest priority iFC in the sequence for processing.

4. Determines if there are any applicable service point triggers in the iFC that match the condition.

5. Will identify the application server address in this particular iFC for forwarding the SIP request, if the service point trigger conditions are met.

6. Will add an identifier in the request to identify the message in subsequent processing and then forward to an application server on the ISC interface.

7. Will continue to apply each filter criteria in the sequence until all are exhausted. If there is a failure to reach to a particular application server, it will perform the default processing specified in the iFC.

Once the S-CSCF invokes the application server, the interaction can be performed in five modes of operation as shown here.

SIP Mode of Operation

Direct processing without an application server

Application server acting as a terminating User Agent or a redirect server

Application server acting as an originating User Agent

Application server acting as a SIP proxy Agent

Application server acting as a B2BUA server performing third-party call control

The simplest case is without an application server intervention. In this case, the S-CSCF acts as a proxy to forward the result to the requested destination. An application server can also choose to use this mode for subsequent requests in the dialog. For instance, an authentication application server may be invoked for the first INVITE request, but for the subsequent INVITEs it may not require any intervention. To enable this case, the application server can avoid placing itself in the Record-Route Header. This will prevent future requests in this dialog to be routed to the application server.

In the case of the AS functioning as a terminating User Agent (UA), the S-CSCF proxies the request to the application server. An example is for a UE requesting a session with a video server to watch streaming video. In the case of the AS functioning as an originating UA, the S-CSCF proxies the request to the requested destination UE. An example is an application server receiving an incoming request from a Web service to talk to a UE.

In a proxy mode, the application server receives the request forwarded by the S-CSCF. It may add, remove, modify header contents, and forward it to the S-CSCF. The S-CSCF then forwards it to the destination.

The B2BUA mode provides the application server with the capability to exercise third-party call control. The AS can operate in two B2BUA modes. As a *routing B2BUA*, the AS receives the SIP request forwarded to it by the S-CSCF. The AS then initiates a new SIP dialog and sends it back to the S-CSCF, which then proxies it to the destination. This mode is typical for a prepaid environment or a 411-like service. In the prepaid scenario, AS receives an incoming call leg requesting for a connection based on credit availability. The AS will then initiate an outbound call leg to the called party. The AS will be responsible for maintaining the two dialogs as a part of the SIP request.

As an *initiating B2BUA*, the AS can generate two or more SIP requests, each as different SIP dialogs. These are logically connected and correlated by the AS. Similar to other cases, the S-CSCF helps to proxy them to their destination. This mode is useful for conferencing servers.

5.8 Service Orchestration and the SCIM

Implementing a solution drawn from standards-based specifications often exposes some gaps that are left to the creativity of the implementers. This classic gap is often perceived as a consequence of difference in thinking between researchers and practitioners.

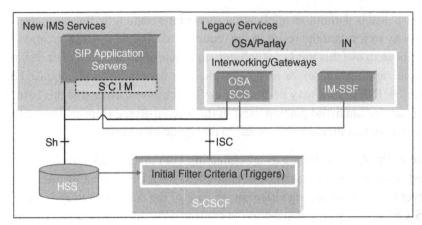

Figure 5.14 The original view of the SCIM.

The reality, however, is that there are three factors contributing to this. In the long-drawn standardization process, it is always a challenge to put on paper a 100-percent solution. The other two factors are advances in technology and changes in market needs by the time the standard comes to see the light of implementation.

The Service Capability Interface Manager (SCIM) is one such example of this classical chasm. This becomes more prominent as we move from proof-of-concept solitary IMS applications toward deployments that rely on multiple service interaction. This does not come as a surprise. The context is new, but the problem is still the unsolved feature interaction reminiscent from the early IN days.

The IMS reference architecture based on best-of-breed telecom solutions continues the similar paradigm, to separate the service plane from the control plane. This offers the extensibility for new service logic, without impacting the CSCF functions. There are two options for the service environment: services that are being built fresh, and traditional or legacy services. We see a new generation of services, which are purely SIP-based or co-exist with legacy services, where the service providers have a significant investment.

As shown in Figure 5.14, IMS clearly outlined the invocation of services from the application servers from the Serving CSCF for next-generation services and for Legacy—an interworking functionality of an IM-SSF for IN services and OSA for Parlay.

The initial standards created the concept where "The application server may contain 'service capability interaction manager (SCIM)' functionality and other application

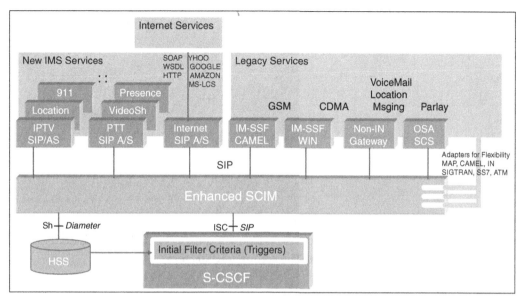

Figure 5.15 The enhanced SCIM view.

servers. The SCIM functionality is an application that performs the role of interaction management. The internal components are represented by the 'dotted boxes' inside the SIP application server. The internal structure of the application server is outside the standards." Since then, however, the SCIM never received any attention from the standards and still is not a bullet item for Release 7. The following drivers have led to an increased interest this interaction management and filling this gap. These are:

- The emergence of blended services

- Limitations of the service triggers (Initial Filter Criteria)

- Inter-working with legacy services

- Converged services with the Internet

Blended Services. One of the benefits IMS promises to offer is a new and unique "User Experience." The reference architecture is intended to provide a rich set of multimedia services. As service providers get a better grasp of monetizing multimedia, the excitement is growing about how to blend services together as opposed to simply bundling them. This paradigm was fairly limited with traditional voice-based telecommunications. The

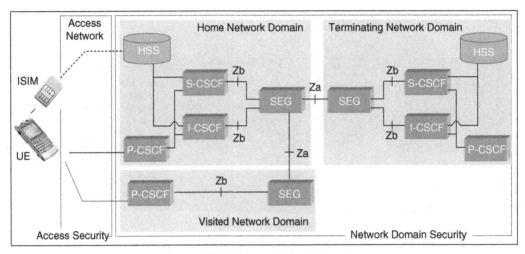

Figure 5.16 The security architecture.

consumers are now expecting Internet-like services on mobile devices. Consider an analogy to a mashup on a google map, which extends the value of a plain satellite image and a street address lookup. Augment that with a push-to-talk to invoke a session to a pizzeria that gives the consumer a new experience for replacing yellow pages. Continuing on the same line to illustrate, blending the potential of IPTV, with location, messaging, and presence services, creates a potential for services beyond Text-based Televoting. What this pushes for is a more intelligent entity to control a set of disparate services. This takes us to the next point, which is to understand better why the current scheme has its own limitations.

Session Control and the iFC. Service Invocation in the IMS is done by a trigger mechanism. The serving S-CSCF obtains the initial filter criteria from the HSS for a registered subscriber. When the subscriber sends a SIP command, INVITE for instance, for the multimedia session one is requesting, the S-CSCF cascades through the iFC and sends the invocation to the necessary application servers based on priority. While the S-CSCF has been enabled to invoke multiple sessions in a sequence, even fork them, it is assumed that once the control is handed over to an application server, it will be the responsibility of the AS to coordinate with other services. This becomes a problem. To implement a service that requires combinational services, such as starting a push-talk session from a google-mashup as described, will require to shift the control between two applications dynamically upon user-control. This also exposes an issue that the iFC can cascade through services, but does not allow interleaving on a dynamic nature.

Legacy Interworking. For the service providers, IMS is a great vehicle for lowered Capital Expenditure (CAPEX) and Operational Expenditure (OPEX), and building differentiated services for future communication networks. However, the challenge for them is that there still exists substantial revenue-generating service infrastructure both within their network and externally (e.g., Messaging, VoiceMail, and Location), that can be effectively harnessed. Adding to it, the initial IMS vision for only CAMEL services (or OSA) to represent a vast legacy infrastructure falls short as well. The CDMA networks are not CAMEL-based but use WIN instead. Apart from roaming and prepaid, few CAMEL services are deployed. So, how would a service provider leverage an investment in, say, a Ringback tone platform or network-based location infrastructure? And orchestration between the new generations of services requires both. For instance, sending an SMS while watching American Idol on IPTV can be done by the click of a remote, but will require the application server to understand that an SMS gateway needs to be used to send the message. So, should IMS push the burden of network technology awareness to Application server platforms, or instead provide a platform to invoke the services in a more agnostic manner?

Internet Services. Given the wide base of a billion Internet users, the expectation of this vast population is to access Internet services on the mobile devices. This again calls for the capability to orchestrate between Internet access and other services. Do IMS Application Servers have centralized access mechanisms to Internet services, or does each have its interface? In other words, does all the presence access get centralized, or do specialized services access these interfaces with their own interfaces?

Given the following discussion, the nature of the problems to be solved now requires the following. This is not truly the SCIM that was conceived of five years ago, but seems to be what is now lacking. In order not to offend the purists, let's call this functional element the *Enhanced SCIM*, or the ESCIM.

The ESCIM can be perceived as an Application Server (AS) platform that has the capability to route, broker, and deliver to specialized application servers.

To the SCSF it simplifies the execution logic of the iFC by interacting with a single entity, and follows the design tenet laid out of pushing control to the application server, which is the ESCIM in this case.

The broad functions of the ESCIM are therefore to provide the following:

- Registration of services

- Invocation of services

- Brokering between services

- Cascading through services by static rules

- Cascading through services through dynamic user interaction

- Routing and load balancing between homogenous application servers

- Providing a user identity for AS-invoked services.

5.9 The Media Plane

The media plane is responsible for the handling of the traffic, which is the various forms of media—voice, data, and video. The media plane elements suitably format and packetize these media forms into Real Time Protocol (RTP) packets. IMS recommends the formats and the codecs for the handling of audio and video streams. This also includes the techniques for speech-enabled and text conversation services.

Digital technology allows us to transform signals from an analog domain into a digital domain. By sampling an audio frequency at certain intervals, the basic form of a digital media for audio can be constructed. By applying algorithms for prediction, interpolation, or adaptation, the sampled signal can yield fidelity to the original signal. These algorithms, sampling frequency, and sampling intervals are some of the parameters that define the codecs—the coding and decoding for the digital signal. Different codecs have been deployed for telephony, speech, and music recording. Some popular codec choices have been the ITU G.711 based on Pulse Code Modulation for a variety of PSTN and SIP applications. The GSM Full Rate is a speech codec based on Linear Predictive Coding.

IMS defines two default codecs for audio in IMS terminals: the Adaptive Multi-rate (AMR) and the AMR-WB (WideBand). The media plane elements are responsible for packetizing and retrieving the audio signals encoded with these codecs into/from the payload in the RTP data, respectively. The AMR-WB encodes audio at a sampling frequency of 16000 Hz in contrast to the 8000 Hz AMR. The AMR-WB is able to deliver a higher quality than the AMR. Both codecs consist of multiple modes, which are codecs working with a different bandwidth.

The AMR codec is a multi-mode codec that supports eight narrow-band speech-encoding modes with bit rates between 4.75 and 12.2 kbps. The sampling frequency used in AMR is 8000 Hz and the speech encoding is performed on 20 ms speech frames. Therefore, each encoded AMR speech frame represents 160 samples of the original speech. Similar to AMR, the AMR-WB codec is also a multi-mode speech codec. The AMR-WB supports nine wideband speech coding modes with respective bit rates ranging from 6.6 to 23.85 kbps. The sampling frequency used in AMR-WB is 16000 Hz, and the speech processing is performed on 20 ms frames. This means that each AMR-WB-encoded frame represents 320 speech samples.

Encoding Video uses a technique to describe the motion video in samples of still images. The encoding of each picture, also referred to as a frame, is digitized into a matrix resolution of a set of pixels. Each pixel is coded to an appropriate color value. A single bit can represent a black and white picture, whereas 8 bits can provide 256 colors. As an example, to show a 256×384 colored frame sampled at 25 frames per second will require a bandwidth of 19 Mbps. Video codecs work to reduce this large overhead by compressing and correlation. The popular codecs are the MPEG, MPEG-2, MPEG-4 from ISO/IEC, and the H.261 and H.263.

IMS identifies the H.263 baseline (profile 0) Level 45 and the Interactive and Streaming wireless profile (profile 3) to be supported as default codec. It also supports the MPEG-4 visual simple profile at Level 3, with certain constraints.

For a text conversation session, where text is sent along with a multimedia conversation, IMS recommends the use of the ITU T.140 codec. For speech-enabled services, IMS recommends the distributed speech recognition (DSR) extended advanced front-end codec. This provides better performance over AMR and AMR-WB speech codecs.

These codecs are specified in terms of the mandatory and optional codecs in the MIME format in the SDP message part. This is used in the SIP invitation for capability negotiation for the media streams to be used. The output from the use of these codecs is presented in payload format in the RTP stream.

These methods and codecs work effectively within the core IMS and the user terminal in a PS domain. Interworking with CS services and VoIP networks introduces some variations in the protocol being used and may require a transcoding function in the media gateway. Both the 3GPP and 3GPP2 standards organizations have adopted the 3G-324M protocol as the primary means of transporting conversational video over mobile networks. This uses similar codecs—AMR and H.263/MPEG4—but the transport is over H.223 instead of RTP.

5.10 Security

The vulnerabilities seen in public IP networks have called for stronger security controls in IMS to secure communications on the lines of the trusted networks deployed in the circuit-switched domain. IMS implements a two-layer security model to enable secure communications. Security mechanisms are built at the access level and at the network domain level. Access Level Security ensures admitting an authorized user to access the network. This requires securing the transport and the mechanism to authenticate the user or the device. Network Domain Security (NDS) refers to the mechanisms that implement a security policy to secure communications between elements within an operator's network and across multiple operator networks.

5.10.1 Security Architecture

As we have learned, the IMS CN architecture adopts an access-independent architecture. To preserve this view, while the IP-CAN can support a security layer at the transport level, IMS additionally adds its own mechanisms to retain this independence. The access level security is achieved by authenticating the UE with a challenge-handshake protocol involving the exchange of the authentication keys and functions. These keys are stored in the UICC or the ISIM module as we observed in Section 5.1. The second aspect of the access security covers the link between the multimedia client in the UE and the first hop, which is the P-CSCF. We will examine these in Section 5.10.2.

The security policy applied within the operator's own network domain or between different network domains is reflected in the NDS. The primary tenet of NDS is to enable a secure connection between the functional elements, both inter-domain and intra-domain. The security mechanisms are applied at the edge of the network by using a specialized element—the Security Gateway (SEG). The SEG interfaces rely on Internet Protocol Security (IPSec) for securing the communication channels. The inter-domain interfaces between the SEGs are identified as the Za reference points. These interfaces are required to support authentication and data integrity protection, and encryption is recommended. This policy is determined by the roaming agreements between the two network domain service providers. The intra-domain interfaces Zb, between the SEGs and the CSCF elements, are required to support both authentication and data integrity protection. However, encryption is optional. The Za and Zb utilize the Ethernet Switched Path (ESP) mode in IPSec, which we saw in Section 4.6.2, for ensuring data integrity protection. The Za typically supports the tunnel mode for inter-SEG communication. IKE is used to negotiate the keys for authentication.

5.10.2 Principles of Network Access Security

Network Access Security in IMS provides two levels of security between the UE and IMS Network.

- Mutual authentication, which is the process of authenticating the subscriber identity with the home network by the exchange key information.

- Secure Link and a security link between the UE and the P-CSCF on the Gm interface.

The mutual authentication is performed by the IMS authentication and key agreement (AKA) scheme. The scheme is implemented with the help of data stored in the IMS Subscriber Identity Module (ISIM) of the UE and the HSS. The subscriber identity—IMS Private Identity (IMPI) used for the authentication—is stored in the ISIM along with the long-term shared key between the HSS and the ISIM. The HSS is responsible for generating keys and challenges. The long-term key in the ISIM and the HSS is associated with the IMPI. The responsibility for executing the process of authentication rests with the S-CSCF.

Following mutual authentication, the UE and P-CSCF negotiate on a set of parameters to set up the security association. The parameters that are exchanged are the encryption algorithm and the integrity algorithm. DES-EDE3-CBC or AES-CBC with a 128-bit key is used for the encryption. HMAC-MD5-96 or HMAC-SHA-1-96 is used for the integrity algorithm.

5.10.3 Principles of Network Domain Security

NDS provides services for

- **Data Integrity or Protection** The data has not been altered in an unauthorized manner.

- **Data Origin Authentication** The data originates from a trusted source.

- **Anti-replay Protection** To protect against the re-use of packets with an encryption mechanism place.

- **Confidentiality** Guaranteeing the information is not made available to unauthorized sources. However, it is limited when flow control is applied.

NDS applies to the control plane within the boundary of the service provider's network domain. NDS does not apply to the user plane. This is in line with what we learned earlier about being able to apply IPSec to Transmission Control Protocol (TCP) and Stream Control Transmission Protocol (SCTP) traffic and not the UDP, which is the transport for the media itself. NDS utilizes the following mechanisms:

- **Security Policy** The set of rules and mechanisms for securing interfaces and providing data protection.

- **Security Domain** The logical or physical definition of the network domain to which a common security policy is being applied.

- **Security Gateway (SEG)** The functional element in a security, which resides at the edge of the security domain that applies.

- **Security Association (SA)** The secure association sets up by the SEG. These are secure IPSec associations and have been negotiated with the IKE.

- **Security Policy Database (SPD)** This contains the set of policies that determine the rules to be applied for all inbound and outbound traffic to the SEGs.

- **Security Association Database (SAD)** Describes the active set of all the security associations and related parameters.

The SEG enforces the inter-domain security policies. These policies apply rules for data protection, filtering, and firewall capabilities. Each SEG is responsible for establishing a security association (SA) with the peer SEG. The security association is encrypted using IPSec in a tunnel mode. The SA uses ESP for the data protection. Establishing an SA is negotiated using the IKE protocol. In order to support bidirectional traffic, the SEG sets up two SAs. One SA handles the incoming signaling and one SA for the outbound.

The SEG maintains two databases. The SEG uses the security policy database to discriminate the traffic on the SA. With the help of the SPD, it determines whether the packets flowing either on the inbound or outbound SA need to be protected by IPSec, or whether they need to be filtered. The security associations' database is used by the SEG as a map of the current traffic set belonging to an SA. Each SA entry is maintained in terms of a security parameter index (SPI), which is an index into the database. It also holds the destination addresses for the SA.

5.10.5 Additional Methods

While the IMS standards continue to fortify the network with this two-tiered model, caution still prevails as hackers are always a step ahead. The threats typically manifest themselves as Denial of Service (DoS), Protocol Fuzzing, Flooding, or Traffic overloads and viruses/malware. There are three additional methods where these security threats can be curtailed. These can be built within a firewall, the border functions (P-CSCF, I-BCF), or with a dedicated security element. The purpose of this function is to detect the following at a minimum. These are based on industry methods and are yet not standardized.

- DoS/Distributed Denial of Service (DDoS) attacks through TCP Synchronous (SYN) flooding and ICMP flooding

- Protocol fuzzing. SIP being a text-based protocol is vulnerable to minor change

- MIME (Multipurpose Internet Mail Extensions) attachment filtering and inspection for virus and malware

- SIP register flooding

- RTP traffic flooding

- Presence updates by spoofed identity

- SIP SPAM

5.11 Chapter Summary

An IMS subscriber is recognized by a unique identity, which enables it to obtain service from the IMS network. This is defined in terms of a public or private user identity. These identities are associated with the UE. Methods such as the GRUU enable a subscriber to maintain multiple UEs with a single public identity. These identities and the profile associated with them are provisioned in the HSS. The subscriber profile data from the

HSS defines the service point triggers that can be invoked on the execution of the initial filter criteria in the S-CSCF during session control. The Authorization and Authentication functions are executed by the S-CSCF and the Application Server based on the service profile and authentication vectors obtained from the HSS.

The heart of the session control is the Leg Control Model, which resides in the S-CSCF. This manages the various inbound and outbound legs in a session, and coordinates the service invocation. Equally important in this function are the P-CSCF and I-CSCF, which route the signaling appropriately to the right S-CSCF.

The Accounting function is implemented as a sophisticated charging function. This can support the different models of offline, online, and flow-based charging. The Policy function has evolved from assuring QoS at the edge of the network and providing bandwidth admission and control. It enables the flow-based charging mechanisms. The Service function is invoked through the ISC interface and comprises the application servers to provide next-generation and converged-legacy services. The SCIM provides the foundation for service interaction and mediation. Security is addressed by IMS at both levels of network access and the network domain.

CHAPTER 6

Putting It All Together

We continue to apply the background of the building blocks of IP Multimedia Subsystem (IMS) and the principles of its operation we have acquired so far. In this chapter we are going to examine the details of how a User Equipment (UE) can invoke services from the IMS core network. While we will focus only on a subset of scenarios, the standards such as TS-23.228 cover a more exhaustive set. We will go through the following sequence in this chapter. What happens when a UE is powered on? How does it determine access to the IMS network? How does the network authenticate and register it? How can the user make or receive a call? How can it invoke services? Finally, how can it operate in a roaming scenario?

6.1 UE IP Connectivity

Obtaining IP connectivity is fundamental in getting access to the Internet Protocol (IP) network. As we have observed, an important tenet of IMS is access network independence. Thus, this procedure is dependent on the IP-CAN being used for obtaining access to the IMS core network. While the methods to obtain connectivity may differ among General Packet Radio Service (GPRS)/Universal Mobile Telecommunication System (UMTS), Code Division Multiple Access (CDMA), Wireless Local Area Network (WLAN), or cable, the steps are similar.

- Determine the network domain from the identity of the user. Authenticate the user with the help of a local Authentication, Authorization, and Accounting (AAA) or an AAA server belonging to the network domain.

- With the assistance of a Dynamic Host Configuration Protocol (DHCP) server, assign an IP address.

- Respond to a request for P-CSCF addresses to obtain admission into the IMS core network.

The P-CSCF addresses for the network being served by the IP-CAN need to be pre-configured. Once the IP-CAN has determined that the UE is eligible for service, it can then allow the UE to discover these addresses. This discovery process allows the UE to be agnostic of a P-CSCF address prior to obtaining IP connectivity.

As you may recall from Chapter 4, we examined two protocols related to IP connectivity: DHCP and GPRS/Evolution-Data Optimized (EV-DO).

While the GPRS/Universal Mobile Telecommunication System (UMTS) IP-CAN provides the framework for establishing the data sessions with PDP context, it relies on DHCP for IP address assignments after having authenticated the user identity with an AAA server. Similarly, 3GPP2-EV-DO and WLAN establish their tunneling methods, authenticate with an AAA server, and obtain the IP addresses from DHCP. In both cases, the DHCP server is also equipped to return the set of addresses for the P-CSCF. In GPRS/UMTS, however, the Gateway GPRS Support Node (GGSN) can be configured to return the addresses for the P-CSCF.

Let's examine the IP connectivity process in the GPRS/UMTS access network, as shown in Figure 6.1. The user connects to the network with its public user identity. The IMS Public User Identity (IMPU) described as a Session Initiation Protocol (SIP) URI or Tel URI, is in the form of user@networkdomain, such as charlie.brown@imsoperator.com. Upon receiving an Activate PDP Request, the Serving GPRS Support Node (SGSN) sends a "Create PDP Context" request to the GGSN, using the virtual APN of imsoperator. The request also includes the username in SIP URI format in the Protocol Configuration Option (PCO) information element (IE).

The GGSN obtains the network domain from the information in the PCO, which corresponds to the real target network on the GGSN. In this example, the GGSN finds imsoperator.com as the domain and directs the session to the appropriate real Access Point Name (APN) for the target network. In this case, the real APN is networkdomain.com.

Next, the GGSN needs to authenticate the user for IP connectivity. If the network domain of the user is the same as the GGSN, it will use the local AAA server to authenticate the IMPU. If the target domain is different, it will request the AAA server of that domain to authenticate the user.

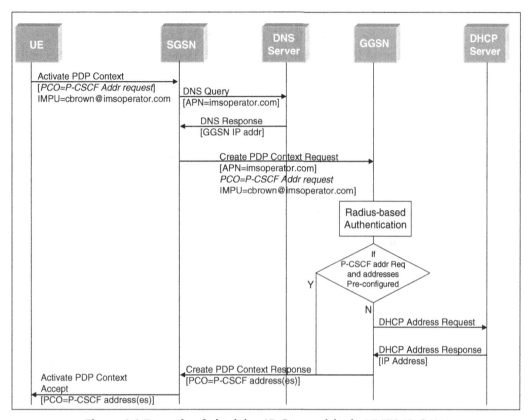

Figure 6.1 Example of obtaining IP Connectivity in UMTS IP-CAN.

Following the authentication, the GGSN obtains the IP address from the DHCP server. The GGSN will also examine the PCO option in the Create PDP Context IE for other requests.

The GGSN can be configured with a set of P-CSCF addresses corresponding to a particular APN. The GGSN will return the preconfigured P-CSCF server addresses for an APN upon request. To obtain these addresses, the UE will need to set the P-CSCF Address Request field of the PCO in the Activate PDP Context Request. This request is forwarded to the GGSN in the Create PDP Context Request by the SGSN. Upon receiving the request, the GGSN returns all the P-CSCF addresses configured in the "P-CSCF Address" field of the PCO.

3GPP requires that the GGSN and P-CSCF must be in the same network (i.e., the visited network). This simplifies the APN resolution for the GGSN to relate with the P-CSCF addresses. Third Generation Partnership Project (3GPP2) does not require the PDSN and the P-CSCF to be in the same network. The PDSN can be in the visited network and the P-CSCF can be in the home network. The DHCP server can be configured to provide addresses for a different network.

If either the GGSN or the DHCP server returns a fully qualified domain name (FQDN) for the P-CSCF, the UE will have to request a Domain Name System (DNS) lookup for resolving the FQDN to an IP address.

6.2 UE Registration and Deregistration

After the UE has obtained an access to the IMS Core Network, the next step is to register itself as an authenticated device, which is authorized to invoke subscribed services. This process of registration with the home network is common for both home network subscribers and roaming subscribers. As shown in Figure 6.2, the functionality for the signaling is common, but the edge access functions are utilized in the visited network for roaming subscribers to register.

Having obtained the address of the P-CSCF, the UE commences the registration process by sending a SIP REGISTER message to the P-CSCF. The message contains the IMPU, IMS Private User Identity (IMPI), the UE IP address, the home network domain, the instance identifier, and the Globally Routable User Agent URI (GRUU) support indication. The P-CSCF uses the home domain network name to determine the address of the I-CSCF of the home network that should service this registration request. It uses a DNS lookup to determine the IP address corresponding to the home network domain. The P-CSCF forwards the REGISTER message to the I-CSCF. The I-CSCF builds a Diameter User Authorization Request (UAR) to request the Home Subscriber Server (HSS) for determining if the registration in the visited network is allowed and if so, obtain the address of the S-CSCF, which should service the registration request. Upon making the determination, the HSS returns the address of the S-CSCF in the User Authorization Answer (UAA), if the user is allowed to register in the network. The I-CSCF then forwards the REGISTER request to the S-CSCF.

The S-CSCF now initiates a challenge-response authentication handshake to authorize the subscriber. The S-CSCF sends a Diameter Multimedia Authentication Request (MAR)

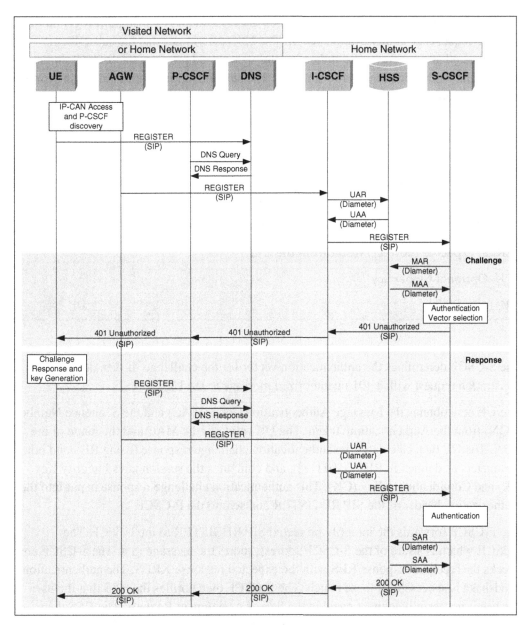

Figure 6.2 Registration.

for the IMPU and IMPI to the HSS. It notifies the HSS that it is serving this registration and also requests an authentication vector. The HSS determines that the subscriber is not registered and generates the authentication vectors and provides the S-CSCF in the Multimedia Authentication Answer (MAA). The authentication is of the following form as defined from 3GPP TS 33.203.

AuthenticationVector = <RANDn, AUTNn, XRESn, CKn, IKn>

RAND: random number used to generate the XRES, CK, IK, and part of the AUTN. Also used to generate the result (RES) at the UE.

AUTN: Authentication token (including MAC and SQN).

XRES: Expected (correct) result from the UE.

CK: Optional Cipher key

IK: Integrity key.

The S-CSCF determines the authentication vector for the challenge. It then denies the registration request with a 401 unauthorized message to the UE.

The UE now obtains the Message Authentication Code (MAC) and the Sequence Number (SQN) from the Authentication Token. The UE validates the MAC and the range of the SQN. The UE then calculates the authentication challenge response (using RES and other parameters as defined in RFC 3310 [18]), and computes the session keys Integrity Key (IK) and Confidentiality Key (CK). This authentication challenge response is put into the Authorization header of the SIP REGISTER and sent to the P-CSCF.

The P-CSCF forwards the integrity protected SIP REGISTER to the I-CSCF. The I-CSCF, which is aware of the S-CSCF address, routes the message to it. The S-CSCF now checks the received response RES with the expected response XRES. The authentication handshake is successful if these match. The S-CSCF then notifies the HSS that the user has been successfully authenticated in the Server Assignment Request. The HSS then returns the user profile to the S-CSCF. The S-CSCF then sends a SIP 200 OK message towards the UE, to complete the registration process.

We examined the scenario for the initial or first-time registration process for the UE. The UE can re-register in situations where the UE capabilities have changed or the UE has been requested for periodic registrations. In this scenario, the UE is already registered with the HSS. When the SIP REGISTER is received by the S-CSCF, it will detect this as a re-registration. The S-CSCF sends the diameter Server Assignment Request (SAR) to indicate the re-registration and obtain any updates to the service profile. The Server-Assignment-Type Attribute Value Pair (AVP) in the SAR indicates Re-registration. Since the UE has already been registered, all signaling coming in is integrity-protected. The authentication challenge is not required for this case.

The de-registration process is the graceful mechanism to terminate the UE's need for services from the IMS network. De-registration can either be initiated by the UE or can be network-initiated. A UE may de-register while powering down or moving into another network. The network may require de-registering a UE for maintenance purposes, network performance, or service contract termination.

In a UE imitated de-registration, the UE sends a SIP REGISTER with the optional SIP Header Expire set to 0. This indicates to the S-CSCF that the UE wishes to de-register for service. The S-CSCF then initiates the diameter Server Assignment Request to indicate the de-registration and obtain any updates to the service profile. The Server-Assignment-Type AVP in the SAR indicates de-registration.

The network-initiated de-registration is triggered differently. It can be triggered either by the S-CSCF or the HSS. The HSS sends a diameter Registration Termination Request to the S-CSCF or vice versa depending on which entity initiates it. The S-CSCF notifies the P-CSCF about this de-registration as well.

The successful UE registration/de-registration also serves as a useful signaling event. Another UE can wish to be notified when the UE is registered in the network and is available for communication. The P-CSCF also maintains a record of the registered UE information. The UE and the P-CSCF subscribe to the registration state event package. This is performed by initiating a SUBSCRIBE request to the S-CSCF. The Event header field is populated with the value "reg" to specify the use of the presence package. The Accept header field is populated with the value "application/reginfo+xml." When the registration or de-registration event takes place, the S-CSCF sends a NOTIFY to the UE or the P-CSCF, which subscribed to this state event package.

6.3 Session Scenarios

6.3.1 UE-UE Calling

Having registered successfully with the IMS Core Network, the UE is ready to request for service. We start our first case of establishing a session with another UE. Let's understand the setup of a UE-UE session in three steps. We first look at the basic protocol required between the two UEs for a message exchange. We then examine the aspects of resource allocation. Finally, we walk through the end-to-end signaling in effect between the elements in the network.

6.3.1.1 The SDP Offer/Answer Protocol

The UE-UE communication requires an agreement on the media and codec resources to be used in the session. This is achieved by the reservation of resources on the UE. In other words, this becomes a *precondition* for the UE-UE session to be established. The support for the preconditions for resource reservation support in SIP is defined and supported in RFCs 3312 and 4032. While it is not mandatory for the originating UE to support preconditions, the standards recommend following the procedure to enable the peer UE to reserve its resources.

The UEs agree on the media and codec resources by using the SDP offer/answer protocol as we see in Figure 6.3. The originating UE initiates the first SDP offer sent in an INVITE message to the terminating UE called "party." The SDP offer contains the set of codecs it is capable of supporting for this session. It also describes the media flows that may be used, the bandwidth required, and their features. When the terminating UE receives this SDP offer, it responds whether the codecs are supported by it. It provides its capability set in the SDP answer in a 183 Session Progress message. The originating UE now makes a determination, which codec can be used for which media flow. For instance, an AMR codec can now be assigned to an audio media stream. The originating UE proposes this in the second SDP offer in a PRACK message to the terminating UE. The terminating UE then acknowledges this offer and answers with its acceptance in a 200 OK message. It also commences its resource reservation for applying the codec and the media stream. When the originating UE receives the acceptance, it applies resource reservation as well. When the resource reservation is completed, it sends the final SDP offer on the media resources that are actually being applied to the terminating UE in an UPDATE message. The called party UE responds back with its acceptance in the 200 OK message. These

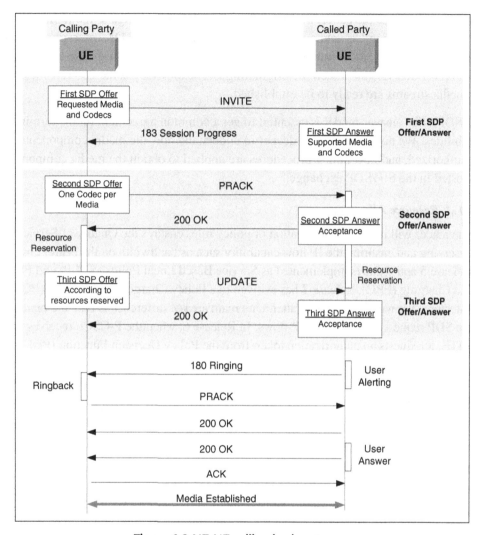

Figure 6.3 UE-UE calling basic setup.

SDP/Offer Response exchanges have now established that the resources required for the call are agreed and available on both the originating and terminating side.

At this stage, the terminating UE may begin optionally alerting the user. It notifies the originating UE with a 180 Ringing message. The originating UE can use this signal to start the ringback to the caller. It sends a Provisional Acknowledge to the terminating UE. The Terminating UE acknowledges the PRACK with a 200 OK message. When the called party answers, the terminating UE sends a final 200 OK in response to the original INVITE for this session. When the originating UE acknowledges the 200 OK message, the media streams are ready to be established.

The SDP offer-answer model is essential to get a common agreement on the terminal capabilities. We now go one level deeper to understand how the media components get authorized, and how policy procedures are applied to obtain the media components requested in the SIP/SDP exchange.

6.3.1.2 Resource Allocation

As the reader will recall, the application of policy rules ensures the Quality of Service (QoS) by reserving and ensuring the IP flow capability such as bandwidth on the bearer channel. In Release 5 and 6 this is implemented as Service Based Local Policy (SBLP) and Flow Based Charging (FBC). Release 7 has evolved to a Policy Charging and Control (PCC) architecture. We will focus on the latter. One main area of difference is how the binding of the SDP request is done to the IP-flows. In Release 6, when the P-CSCF receives an INVITE, it requests an authorization token from the Policy Decision Function (PDF). The P-CSCF sends this Authorization token in a P-Multimedia-Authorization header to the UE. The UE would then use this token along with the IP-flow IDs in its PDP context activation/ modification request to the GGSN. The token helped the GGSN to resolve the address of the PDF and it could obtain the bearer authorization from the PDF. Release 7 does away with the concept of the token, but allows it to support for backward compatibility.

In a Release 7 scenario as shown in Figure 6.4, when the P-CSCF on the originating side receives the first SDP offer in the SIP INVITE, it identifies the connection information for the downlink. This includes the address and port of the IP flows for the stream between the network to the UE. When the terminating P-CSCF receives the INVITE, it identifies the connection information for the uplink. This includes the address and port of the IP flows for the stream between the UE to the network. When the terminating P-CSCF receives the SDP answer, it identifies the downlink flows for the terminating UE. It then creates a Diameter AA Request command to send to the PCRF. It sets the Media-Component-Description AVP with the information it has received from the SDP offer-answer. This includes the requested media type and the maximum requested bandwidth for the uplink and downlink. When the PCRF receives the Diameter AAR, it identifies the

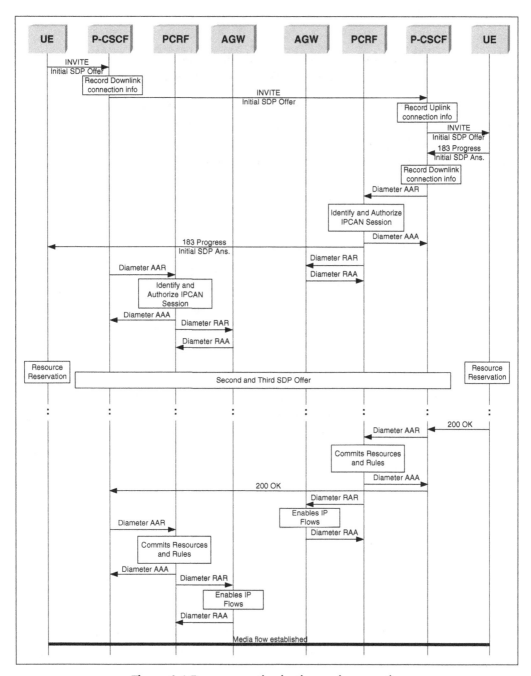

Figure 6.4 Resource authorization and reservation.

IP flows bound to the session or the bearer. It authorizes these rules by installing these and communicating to the Access Gateway (AGW) with the Diameter RA Request.

The Originating P-CSCF will do the same on its side for the resource authorization upon receiving the first SDP answer. Subsequently, the session establishment continues to progress and the called party answers. At this stage, when the P-CSCF receives the 200 OK, it is ready to direct the P-CSCF to enable the IP-flows. It sends an AA Request to the PCRF with the Flow-Status AVP-enabled. The PCRF then directs the AGW with a RAR to indicate establishment of the bearer. The media flows at this stage are ready to be established.

6.3.1.3 SIP Header Functions

Before we go to the final step, let's examine the role of some important SIP headers and the operations on them. The P-Charging-Function-Addresses header helps to identify the instance of the charging servers in the network. The address of the online and offline charging servers are encoded in this header to allow the routing of the message.

```
P-Charging-Function-Addresses: ccf = 172.20.10.1;
ecf = 172.20.10.8
```

To correlate the charging information, the IMS Charging Identifier (ICID) is exchanged between the elements participating in a particular session. The P-Charging-Vector header is used to transport the ICID. This header is populated by the P-CSCF, and also contains additional information. The access network charging identifier is used to correlate the IMS charging data with the access network charging data. The Inter-Operator Identifier (IOI) is inserted by the S-CSCF for network use. It is removed by the P-CSCF when the request goes outside the network.

```
P-Charging-Vector: icid-value = "831Crtp634+1467219834e";
orig-ioi=globus.com
```

During the process of establishing a session or a transaction, an element may opt to remain in the path of subsequent routing or opt out. The Record-Route header is inserted by an element to ensure that the subsequent requests or responses for a particular session must be routed through it. It can delete the header when it chooses to opt out. As we will see, the P-CSCF will insert the Record-Route header to enable subsequent messages to be routed through it. The I-CSCF, on the other hand, will remove the Record-Route after it has located the S-CSCF and routed the message to it.

The P-Preferred-Identity and the P-Asserted-Header headers are used in the process of establishing a trusted network. The P-Preferred-Identity header field is sent by the UE to a trusted P-CSCF. It carries the user's identity such as the IMPU, which it wishes to be used for the P-Asserted-Header field value that the trusted element will insert.

6.3.1.4 Message Exchange

With this background, let's explore the complete signaling path. We look at the significant elements in the SIP and SDP message that are exchanged to set up the UE-UE session. The originating UE initiates the session with the INVITE message directed to the P-CSCF in the serving network. The INVITE message contains the following information in its header. The Request-URI contains the IMPU of the terminating UE. The Via field is set to the IP address or the FQDN of the UE. The Route Header contains the address of the P-CSCF. Since the message comes in over the air and requires compression, the description will include the SigComp indication. The P-Preferred-Identity holds the preferred IMPU of the calling party to be used for the session. The P-Access-Network-Info describes the details of the serving access network. The Supported header contains precondition 100rel to indicate the support for the precondition mechanism and reliable provisional responses, respectively.

When the P-CSCF receives the INVITE, it acknowledges with a 100 Trying provisional response to the UE. The P-CSCF adds its entry to the Record-Route and Via Header. It removes the SigComp indicator, as it is not required for the network. The P-CSCF also replaces the P-Preferred-Identity header with the P-Asserted-Identity containing the IMPU of the calling party. The P-CSCF also inserts the P-Charging-Vector with the ICID parameters. The P-CSCF now forwards the message to the S-CSCF of the home network of the calling party.

When the S-CSCF receives the message, it acknowledges with a 100 Trying provisional response to the UE. The S-CSCF then validates the service profile of this subscriber and determines the Initial Filter Criteria for this subscriber. Since the message may now go to a different operator network, the S-CSCF adds its network identifier to the IOI of the P-Charging-Vector. The S-CSCF now forwards the INVITE to the I-CSCF of the called party's home network. The I-CSCF sends a 100 Trying response to the S-CSCF.

The I-CSCF must now determine which S-CSCF in the called party's home network should handle the session. It constructs a diameter Location Information Request. It sets the Public-Identity AVP to the IMPU that has been received in the Request-URI and sends it to the HSS. The HSS returns the server name of the S-CSCF. The I-CSCF then forwards the INVITE to the S-CSCF in the called party's home network. The S-CSCF acknowledges the message with a 100 Trying and forwards the INVITE to the P-CSCF of the called party's serving network.

The P-CSCF now initiates service policy procedures and bandwidth reservation for the call. The P-CSCF communicates with the Policy and Charging Rules Function (PCRF) about the resources required for the establishment of the call, as described previously.

The P-CSCF now adds its agreed port number in the Via header to the Record-Route. Also, since now the message will go over the air, it sets the compression option to SigComp. (In a Release 6 scenario, it will set the P-Media-Authorization to the authorization token.) It now forwards the INVITE to the UE. As outlined earlier in this section, the UE has received the first SDP offer in the INVITE. It returns a 183 Provisional response to indicate the media stream capabilities supported by it.

Upon receiving the 183 response, the P-CSCF now requests for the authorization of the bearer resources from the PCRF. It removes its port from the Record-Route header and the SigComp option, and forwards the message to the S-CSCF of the called party's home network. The S-CSCF in turn forwards the message to the calling party's S-CSCF, which in turn sends it to the P-CSCF in the serving network. The P-CSCF requests for authorizing the media resources. It adds its port in the Record-Route along with the compression option set. It then forwards the 183 provisional response to the calling party UE. The calling party UE now sends a PRACK for the next SDP offer. If the called party answers with one codec per media stream, no SDP shall be sent; otherwise, the SDP will contain the preferred codec by the calling party UE.

The PRACK is now sent to the called party, routed through the calling-party's visited network P-CSCF, the home network S-CSCF, the called party's S-CSCF, and the visited network P-CSCF. A similar return path takes place for the 200 OK and the final SDP offer with the UPDATE message.

The called party UE must ensure that resource reservation on its end has been successfully completed and the resource reservation on the calling party has been completed as well. The receipt of the UPDATE message indicates that it has been completed on the originating side. Only after successful completion of these two preconditions can it continue with the session to initiate alerting the user and indicate to the calling party UE that the ringing has begun. The 180 Ringing message is routed via the P-CSCF and S-CSCF.

The originating UE now commences the ringback to the calling party and sends a PRACK to the UE. This is routed through the P-CSCF and S-CSCFs. The called party UE responds with a 200 OK.

Having successfully allocated resources and alerted the users on both ends, the called party acknowledges a 200 OK to the INVITE that was sent by the calling party to initiate the session. The P-CSCF now sends a commit to the PCRF to enable the resources that were reserved. The calling party sends an acknowledgement (ACK) and the media streams are ready for connection.

6.3.2 UE-PSTN Calling

We will now explore a similar scenario, where the calling party UE wishes to establish a session with an endpoint in the circuit-switched network. We consider the similar scenario where the calling party is roaming in a visited network. The basis for the session setup is quite similar to the previous UE-UE call model (Figure 6.5). The reader can appreciate and draw a parallel with the flow of control from the S-CSCF to the Border Gateway Control Function (BGCF) and Media Gateway Control Function (MGCF) similar to the I-CSCF and S-CSCF of the called party in the UE-UE session.

The reader may also recall from Chapter 1 about the role of the BGCF. When the S-CSCF determines the called party is a PSTN destination, it forwards the request to a local BGCF. Based on further analysis of the destination address, and on agreements between operators for PSTN termination, the BGCF will either select a local MGCF to perform the termination, or will forward the request to a BGCF in another operator's network, which will select the MGCF to perform the termination. In this section, we will consider the case with the local MGCF. The flows are similar.

The originating UE initiates the session with the INVITE message directed to the P-CSCF in the serving network. The INVITE message contains the following information in its header. The Request-URI contains the identity of the terminating UE as a tel-URI. The Via field is set to the IP address or the FQDN of the UE. The Route Header contains the address of the P-CSCF. Since the message comes in over the air and requires

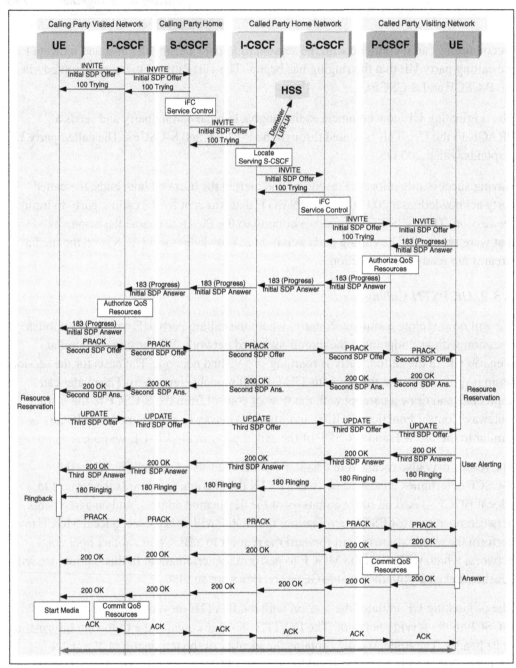

Figure 6.5 UE to UE calling.

compression, the description will include the SigComp indication. The P-Preferred-Identity holds the preferred IMPU of the calling party to be used for the session, which is in the tel-URI form. The P-Access-Network-Info describes the details of the serving access network. Preconditions are required, hence the Require header is set to preconditions. The Supported header contains 100rel to indicate the support for the reliable provisional responses.

When the P-CSCF receives the INVITE, it acknowledges with a 100 Trying provisional response to the UE. The P-CSCF adds its entry to the Record-Route and Via header. It removes the SigComp indicator, as it is not required for the network. The P-CSCF also replaces the P-Preferred-Identity header with the P-Asserted-Identity containing the tel-URI IMPU of the calling party. The P-CSCF also inserts the P-Charging-Vector with the ICID parameters. The P-CSCF now forwards the message to the S-CSCF of the home network of the calling party.

When the S-CSCF receives the message it acknowledges with a 100 Trying provisional response to the UE. The S-CSCF then validates the service profile of this subscriber and determines the Initial Filter Criteria for this subscriber. The S-CSCF makes an analysis that the destination is a PSTN number and has to be routed to the BCGF. Since the message could potentially go to a different operator network, the S-CSCF adds its network identifier to the IOI of the P-Charging-Vector. The S-CSCF inserts the P-Charging-Function-Addresses header to provide the charging function addresses to the BGCF.

When the BGCF receives the INVITE, it sends a 100 Trying message to the S-CSCF. The BGCF analyzes the termination address and makes a determination whether to allocate a local MGCF or an MGCF in a different network. Note that the BGCF does not add itself to the Record-Route header, as it has no need to remain in the signaling path once the session is established The BGCF then routes the INVITE message accordingly. In our scenario, it routes it to the local MGCF. Upon receipt of the INVITE, the MGCF sends a 100 Trying message to the BGCF.

The MGCF now commences the reservation of its connection resources and configuration of remote resources. The MGCF can select the IM Media Gateway (IM-MGW) for the bearer before the signaling starts, or during bearer negotiation. In the previous scenario, we consider the former case. The MGCF sends the H.248 ADD.request message to create the context and termination. It configures the local and remote IP address, the port, and the codecs that will be used in this communication.

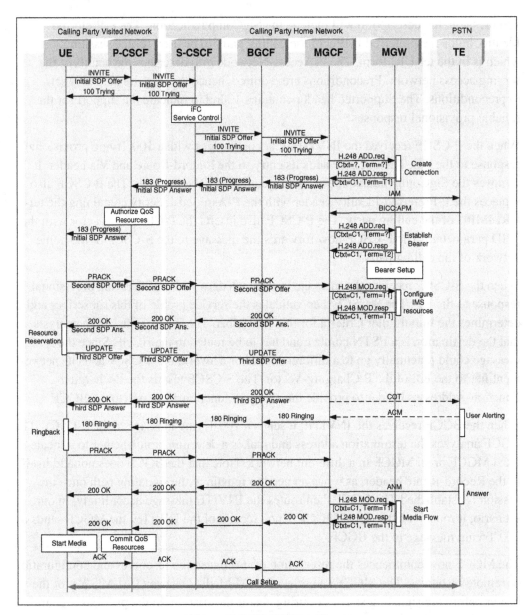

Figure 6.6 UE-PSTN calling.

The MGCF then returns its media capabilities, which is the IP-address and port for the media-stream and the codec(s) in the SDP answer. It provides this in the 183 Progress message to the BGCF, which gets routed to the UE. The MGCF initiates the preparation and establishment of the bearer. It sends out a Bearer Independent Call Control (BICC) or an ISDN User Part (ISUP) Initial Address Message (IAM) to the terminating equipment. Based on the BICC Application Transport Mechanism (APM) response, it will direct seizure of the Circuit Switched (CS) trunk and establish a forward path for the bearer.

When the MGCF receives the PRACK originated from the UE, it reserves the resources on its side of the IMS CN and responds with a 200 OK message toward the UE. The UE performs resource reservation on receiving the message, and sends an UPDATE toward the MGCF. It responds with a 200 OK message. If continuity tones for CS networks are required, it sends a COT optionally to the TE. When the called party starts initiating an alerting, it responds with an Answer complete message (ACM). The MGCF then returns a 180 ringing message toward the UE, to indicate that user alerting has begun. The UE acknowledges with a PRACK commencing ringback on its end. When the called party answers, the PSTN sends an Answer Message (ANM). The MGCF now requests the MGW for a two-way bearer connection. It responds with a 200 OK(INVITE) message to the UE. The UE commences the media flow on its side, and the P-CSCF commits to enable the IP-flows at the IP-CAN. The full bearer path is now ready to be set up.

6.3.3 PSTN-UE Calling

In continuation of the PSTN scenario, let us now follow through the setup of a session originated from the CS network to a UE. As we observe in Figure 6.7, there is no requirement for a BGCF for an inbound call into the IMS CN from the CS network. The inbound CS signaling terminates at the MGCF and the MGCF initiates the session to the UE.

The CS Network establishes a bearer path to the MGW, and signals to the MGCF with a IAM message, giving the trunk identity, destination information, and optionally the continuity indication. An inbound ISUP or BICC IAM message is received by the MGCF. The MGCF then initiates H.248 signaling to the IM-MGW to establish the bearer, and commences the session to the UE. The MGCF sends an H.248 ADD.req to the IM-MGW to create a context and add a termination for the IMS connection point. The MGCF also initiates a SIP INVITE with the initial SDP offer. It sets the Request-URI to tel-URI that has been received in the IAM called party information element. It sets

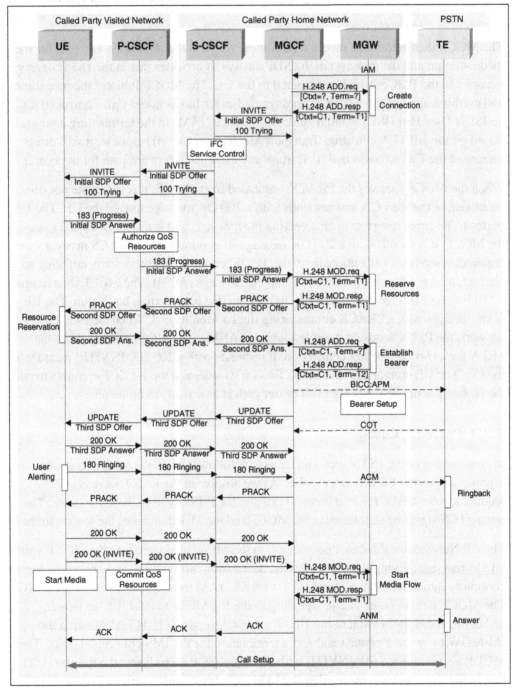

Figure 6.7 PSTN-UE calling.

the P-Asserted-Identity to the tel-URI of the PSTN calling party. The Require Header is set to Preconditions, which are to be supported. The Via holds the IP address or FQDN of the MGCF. The MGCF also inserts the P-Charging-Vector header and sets the ICID parameters. It then sends the message to the S-CSCF.

Upon receipt of the message, the S-CSCF sends back a 100 Trying message. The S-CSCF determines the route to the UE. From the registration information of the UE, it is able to determine the address of the serving P-CSCF. The P-CSCF routes the message to the UE. The UE then returns its media capabilities in the SDP answer in the 183 Progress message via the P-CSCF S-CSCF signaling path to the MGCP. The MGCP then reserves the bearer resources on its side by directing the IM-MGW with a H248 MOD.req message. It returns the selected codec and media flows in the PRACK to the UE. The UE now begins resource reservation on its side. When the MGCP gets the 200 OK message from the UE, it commences the seizure of the bearer. It sends a BICC APN to establish the bearer, and responds to continuity testing if supported.

When the user alerting commences on the called party UE, the UE notifies with a 180 Ringing message. The MGCF then initiates an Address Complete Message (ACM) to the Public Services Telephone Network (PSTN) to indicate the ringback on the calling party and notifies the UE with a PRACK. When the called party answers, the UE sends a 200 OK INVITE message and begins the media-flows on its side. Upon receiving the notification, the MGCF indicates an ANM to the PSTN. It signals to the IM-MGW to enable a bidirectional flow with the H.248 MOD.req message. It returns an ACK to the UE and the full bearer path is ready to be set up.

6.3.4 UE Service Invocation

We now examine our next example of how a UE invokes a service from an application server as illustrated in Figure 6.8. We continue with our case where the UE is in a visited network and initiates an INVITE session to a PSI hosted on an application server.

The UE begins with an INVITE with the Request-URI containing the address of the PSI the UE wants to establish a session with. It forwards the message to the P-CSCF, which then sends it to the I-CSCF in the home network.

When the I-CSCF receives the INVITE message, it checks the Request-URI. If it contains a pres: or an im: URI, it will translate that to a public user identity. It removes the

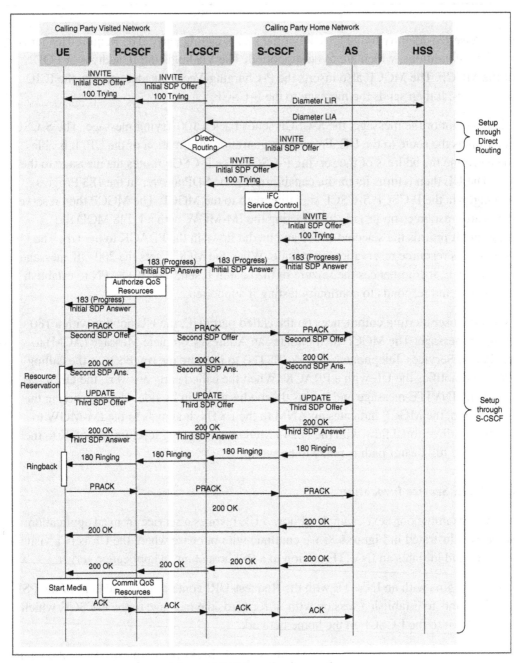

Figure 6.8 UE service invocation.

Route-Header. If the domain name in the Request-URI matches the known Public Service Identity (PSI) sub-domains in the I-CSCF, then it resolves the Request-URI by a DNS lookup to determine the address of the AS hosting the PSI. Otherwise, the I-CSCF will send a Diameter Location Information Request (LIR) to the HSS for the PSI derived from the Request-URI. If the Location Information Answer (LIA) returned from the HSS contains the address of an application server, it will then forward the request to the AS directly, which is hosting the PSI. If the LIA returns the address of the S-SCSF assigned to handle the request, will forward the INVITE to the S-SCSF. In either case, it will retain the value of the ICID in the P-Charging-Vector.

If the UE is trying to establish a call-related session, the message flow thereafter is similar to the session establishment as we followed earlier.

6.3.5 PSI-UE Calling

We now consider a case where a Web user wants to establish a session with an IMS UE. Web services may or may not support a SIP protocol (e.g., a skype). This requires an application server that can host this session as a public service identity (PSI). The Application server would be capable of supporting the necessary Web protocol, codec, and the inter-working with the SIP/SDP for the IMS Service Control (ISC) interface.

When the PSI requests to initiate a session, the Application Server (AS) creates an INVITE message with the following. It inserts the Request-URI to the IMPU of the UE being invited for the session. It sets the P-Asserted-Identity to the PSI. It requests the HSS with a Diameter Location Information Request to obtain the address of the S-CSCF serving the PSI. It sets up the SDP with the media-flows and codec capabilities supported by it. It also sets the P-Charging-Identifier with the ICID parameter populated. The AS then forwards the INVITE with the initial offer to the S-CSCF.

The S-CSCF builds an ordered list of the iFC based on the PSI in the P-Asserted-Identity. It determines if the content of the SDP matches the subscribed services of the called party UE. It removes its SIP URI from the Route Header. It determines the P-CSCF serving the UE and forwards the INVITE to the P-CSCF.

The session setup goes through the common process of the SDP Offer-Answer for the resource reservation, followed by ringing and answer.

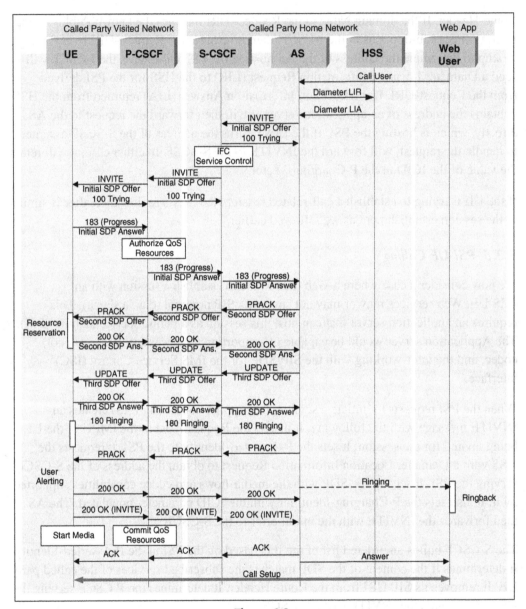

Figure 6.9

6.4 Chapter Summary

In this chapter, we walked through the basic communication scenarios for a UE. We explored a small but fundamental set. The first step is for a UE to obtain IP connectivity to the IMS core network and discover the P-CSCF. This process is dependent on the IP-CAN being used. We examined the GPRS/UMTS case of how the PDP context gets established and the mobile is able to get an IP address through DHCP and DNS support. The UE is able to reach the Proxy-CSCF in the serving network. This could either be a visited or a home network. Subsequently, the UE must register itself with the core network. This includes the authentication process. We also looked at the re-registration and the de-registration process.

We explored the session scenarios subsequently. These included origination and terminating sessions between two IMS UEs, an IMS UE, a PSTN terminal, and an IMS UE and a PSI. We learned the concept of preconditions, which is the agreement of media and codec resources to be used in a UE-UE session. We saw the applicability of the SDP Offer/Answer model as a building block in the message interactions. We applied the concepts of the iFC, service invocation, and media interworking we learned earlier, in these session flows. In the next chapter we will focus on how the UEs can invoke multimedia services.

CHAPTER 7

Services Delivered by IMS

What really makes IP Multimedia Subsystem (IMS) exciting is its capability to deliver a diverse set of services. Because of this, IMS is viewed as a platform in the network, which can scale from basic telephony services to next-generation multimedia services. We are going to examine the nature of services that are possible with IMS, and then explore them in detail. The initial trend of services with IMS has been to host the contemporary applications such as presence, messaging, and Push to Talk. The true potential of IMS is to deliver converged, combinational, and blended services, which can offer a truly rich experience to the user.

7.1 Converged Services

Access to the network is the foremost and fundamental service delivered by a network service provider; however, it is basic in nature. Service providers have benefited more with services that can deliver value or add value beyond simple access. Circuit-switched networks provide this value with four types of services: circuit teleservices, bearer services, supplementary services, and value-added services. Examples of circuit teleservices include telephony, emergency calling, and Short Messaging Service (SMS); supplementary services include Caller ID, Call Forward, and value-added services such as prepaid and ringback tone. These have been the cornerstones of wireless networks.

Packet-switched networks have provided a greater value by the Internet Protocol (IP) bearer services. These services can be call-related or non-call-related. IMS networks rely on the IP bearer services of packet-switched networks to deliver a rich set of multimedia session-related services. While we will examine the nature of multimedia services shortly, the true value of services IMS delivers is seen by its capability as a convergence platform. This allows extending the value of circuit-switched services in the packet-switched domain.

7.1.1 Multimedia Services

A *multimedia service* is a combination of two or more media components such as voice, data, video, and still image, in a single session delivered between two or more parties. Multimedia services can be classified as interactive or distribution services. User interaction defines the behavior of interactive services. Distribution services, on the other hand, are broadcast-based. There are three categories of interactive services: conversational, messaging, and retrieval.

Conversational services are based on user-to-user interaction. To maintain parity with human perception, these services are usually bi-directional with low end-to-end delays. Media synchronization is required to avoid low delay variation such that there are no perceived gaps or jitter. Real-time responses are required and there is no need for any store and forward of data. Examples of conversational services include video telephony and video conferencing.

Messaging services also provide user-to-user communication. However, it requires a store and forward mechanism, such as a mailbox or a mail server. Multimedia messaging services can support video, audio, text, and still images, individually or in combination.

Retrieval services provide the mechanism for a user to obtain information stored from a repository. The user can access audio, video, text, and still-image on-demand.

Multimedia Distribution services are based on broadcast services. These can be delivered with or without user presentation control. Distribution services without user control are simple broadcast services such as audio or video. The multimedia information is transmitted from a central point. The user has no ability to control the timing or delivery of the content. To support user control, the distribution services broadcast content by using sequence numbering and repetitive transmission. This allows the user to control the presentation of content in terms of sequence and delivery of the content.

7.1.2 Service Paradigms

Four service paradigms are often mentioned in the context of next-generation networks: bundled services, converged services, combinational services, and blended services. *Bundled services* refer to the ability of a service provider to provide a group of access-based services such as voice, video, and data as a package. Bundling of services is a function that can be done by the operator without the use of IMS. The other three services paradigms are of more interest to us in using IMS.

Telecommunication services traditionally focused on delivering access to the network. Since then, services have evolved to become application-focused rather than being pure access-centric. This delivers value not just to the user by a new experience, but to the network operator as well as by the ability to monetize. Applications tend to have "stickiness" in their user base and prevent churn. As users get exposed to applications across heterogeneous networks and different devices, their expectation consequently rises to see common applications. This is one of the drivers for convergence. *Converged services* aim to deliver a unified experience regardless of the network or the device.

In the current stage of deploying IMS, converged services today typically require delivering uniform voice, data, and video services as they are provided across different networks and devices. These are typically circuit-switched telephony services such as Calling Party Name, 911, 411, and ringback tones; delivering streaming video, video-on-demand, and multipoint-conferencing video service; and delivering browser-based applications on the data plane. Converged services focus on integrating with a legacy or a traditional network via interworking.

Combinational services refer to the method to deliver a multimedia service as a combination of a circuit-switched call and a packet-switched session. A combinational service can be created by adding one or more IP multimedia component(s) to a Circuit Switched (CS) call. The CS and IMS components are established between the same participants. The resulting service can either be a combinational call or a combinational session. A circuit-switched voice call that can be enhanced with a multimedia component by adding an IMS session between two User Endpoints (UEs) forms a combinational call. Likewise, in a combinational session, an IMS session in progress between two users is enhanced by adding a circuit-switched base. Video sharing and other collaborative sharing of applications and whiteboard are examples of combinational services. Combinational services are an efficient method to utilize the existing CS infrastructure with IMS to deliver a multimedia experience.

Blended services provide the capability to have the concurrent use of voice, video, and data features in one session and to move seamlessly between functions provided by these multimedia applications. Examples of blended services are:

- Click to dial from a map or a video stream.

- Change an instant message session to a voice call or video conference call.

- Switch from a voice call to a multiparty video conferencing session.

- Browse the Web while on voice/conference call, with the potential to add Web sharing.

- Send/receive voice mail, video mail, or e-mail while on a voice call.

7.2 Next-Gen Consumer Services

7.2.1 Presence

The concept of *presence* is not new to wireless cellular networks. These networks have utilized the presence of a user within a network to provide the function of mobility between a home and a visited network. Presence, however, as it is understood today, refers to a paradigm of a network user entity announcing its availability within a network to a set of interested entities. This concept of presence differs from the presence in the established mobility model in two ways. Presence, as it is defined, is in the control of the user. It reflects a user-directed status. Second, the value of presence is to entities other than just the network itself.

A similar term that is used for the situation of the user is "Location." This should not be confused with presence. *Location* refers to the exact positioning information typically related to the geographic area about the user.

The next related term is *context*, which refers to any information that can be used to characterize the situation of an entity. Context-awareness can be either derived from the presence or location information. Presence may be used to invoke an interaction (e.g., instant messaging). Context refers to the capability to direct the behavior of the interaction based on the situation of the entity. For instance, a context-aware application may localize the news delivery to a presence in a roaming network.

While the IMS standards do not focus on methods of context-awareness, they do define presence as an important model to enable converged applications. The essence of presence as described in the TS 122.141 is for the IMS network to be able to support a central model of a presence service, which can aggregate and disseminate the information to other networks, including inter-working with the Internet applications. This provides a backbone for presence-enabled services for voice, data, video, and multimedia, as shown in Figure 7.1.

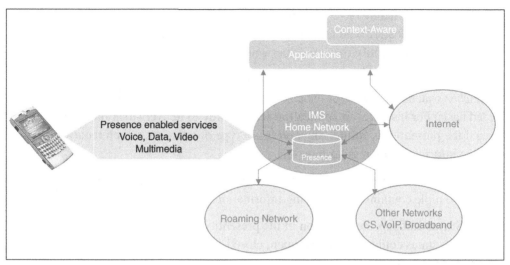

Figure 7.1 Presence-enabled services.

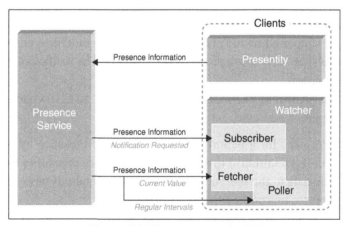

Figure 7.2 The presence model.

The presence model in IMS has been adopted from the Internet Engineering Task Force (IETF) (ref RFC 2778). The presence model defines a client-service paradigm as shown in Figure 7.2. Clients referred to as the *presentities* make their presence information available to the presence service to be stored and distributed. The clients requesting this presence information from the presence service are referred to as the *watchers*.

Depending on their nature of request, watchers are of two types. A *subscriber* watcher requests for a notification when there is a state change in the presence information. The presence service will notify the subscriber watcher whenever the presentity changes its state from busy to available.

The *fetcher* watcher requests for the current value of the presence information on an as-needed basis. For instance, an application server ready to deliver stock quotes may check to see if the presentity is online. The *Poller* is a type of a fetcher, which requests the presence information on regular intervals.

Presence information for each presentity is organized as set of *presence tuples*. A presence tuple contains the following information.

- **Status** Reflects the condition of the presentity and its willingness for interaction. The status could typically be open, closed, online, offline, busy, away, do not disturb, and so forth.

- **Communication Address** The means and the address to reach the presentity.

  o **Communication Means** This reflects the mechanism of interaction, which can be supported by the presentity. The means could be a combination of service type such as telephony, SMS, and IM. The media could be audio, video, or text.
  o **Contact Address** The identifier such as a URI, an E.164 address, or an IM instant inbox address.

Support for Rich Presence Information data (RPID) and the Presence Information Data Format (PIDF) is an IETF initiative to add service and device capability and geographic location and privacy data to the presence tuple. The IMS standards have not formally incorporated support for this at the time of writing.

We now see how the IMS architecture applies the presence model, as shown in Figure 7.3. The *principal* is the user of the presence service and could be a human, group, or applications. A principal is associated with one or more presentities or watchers. There are three types of presentities identified:

- The Presence User Agent (PUA), which is a component of the UE

- The Presence Network Agent (PNA), which could be resident on any of the IMS CN elements

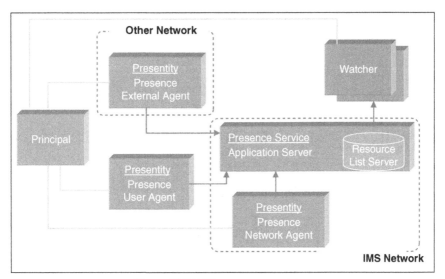

Figure 7.3 IMS presence entities.

- The Presence External Agent (PEA), which may access the presence service from a different service provider network

The presence service is implemented as an application server, accessible via the SIP interface. IMS also defines a resource list server for group list management, supporting buddy list applications.

The interaction between the presence entities takes place with a SIP "pub-sub" model as described in Figure 7.4. The watcher sends a SIP SUBSCRIBE message to the presence server to subscribe to the presence of a particular presentity. The presence server makes a determination if the watcher is authorized for this service, and is allowed by the presentity to know about its presence. The presence server returns a 200 OK message if it is successful. The presence server also notifies the watcher with the current presence status of the presentity in a NOTIFY message. The watcher acknowledges with a 200 OK message.

The presentity uses the SIP PUBLISH to announce its presence status. It may either be registering for the service and announce its availability for communication, or it may use it to update its status to busy or unavailable. Upon receiving a PUBLISH, the presence server authorizes the presentity for the service and, if successful, records its status and sends the 200 OK to the presentity.

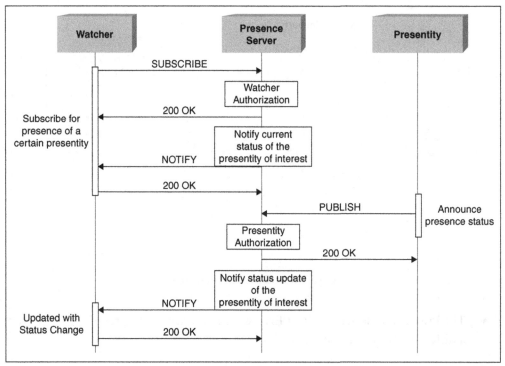

Figure 7.4 The generic SIP model for presence.

The presence server then looks up all watchers that have subscribed for a notification of this particular presentity's status. It then forwards a NOTIFY message to the watcher notifying of a status change of the presentity. The watcher would acknowledge the status change with a 200 OK message.

Applying the generic model to the IMS network is fairly straightforward. We consider a case with the presentity and a watcher being two PUAs in the UE. For this case, the presentity UE is roaming in a visited network and the watcher is registered in the home network.

The watcher UE generates a SUBSCRIBE request containing the presence event it wishes to be notified of, together with an indication of the length of time this periodic subscription should last and the support for partial notification. It sets the Request-URI to the IMPU of the presentity of whose presence it wishes to know, and the Event header is set to presence. It then sends a SUBSCRIBE message to the P-CSCF in the home network. The P-CSCF forwards the SUBSCRIBE message to the serving S-CSCF for this UE as it was established during the registration, and adds a Route header.

The S-CSCF evaluates the Initial Filter Criteria (iFC) for the SUSBCRIBE message received, and forwards the SUBSCRIBE to the presence server. The presence server authorizes the subscription, makes sure all privacy checks are met, and sends a 200 OK message to the S-CSCF. If they fail, it would send a 4xx message. The S-CSCF then returns the 200 OK message to the watcher UE via the P-CSCF. The presence server then initiates a NOTIFY to notify the watcher UE about the current status of the presentity UE. Based on the Accept Header message sent by the watcher, it can communicate partial notification of the current status. It sends the NOTIFY to the S-CSCF and the S-CSCF forwards it to the watcher UE. The watcher UE acknowledges with a 200 OK message.

At some point, the presentity UE sends a PUBLISH message to the P-CSCF in its visited network. The UE sets the Request-URI to the IMS Public Identity (IMPU) of the PUA. It sets the To header to the same as well. The P-CSCF determines the home network of the UE and forwards the PUBLISH to the I-CSCF in the home network of the UE. The I-CSCF requests the user location from the Home Subscriber Server (HSS) by sending a Diameter Location Information Request (LIR). It receives the address of the serving S-CSCF. The S-CSCF determines the address of the presence server by validating through its iFC, and sends the message to the presence server.

The presence server authorizes the presentity UE and sends a 200 OK message to the S-CSCF. The S-CSCF then forwards the message to the presentity UE.

The presence server then determines the watcher UE list subscribed for the presence of this presentity UE and sends them a NOTIFY message (see Figure 7.5).

While we examined the SIP flows required for the presence exchange, the presence information itself is exchanged in the XML Configuration Access Protocol (XCAP), which we recall from Section 4.8.1. XCAP is used to store, alter, and delete data related to the presence service. XCAP is designed according to the HyperText Transfer Protocol (HTTP) framework, and uses the HTTP methods PUT, GET, and DELETE for communication over the Ut reference point. The general information that can be manipulated is user groups, subscription authorization policy, resource lists, and hard state presence publication. The advantage of using XCAP as a protocol is to enable a uniform interface with both the Presence User Agent (PUA) and the Presence External

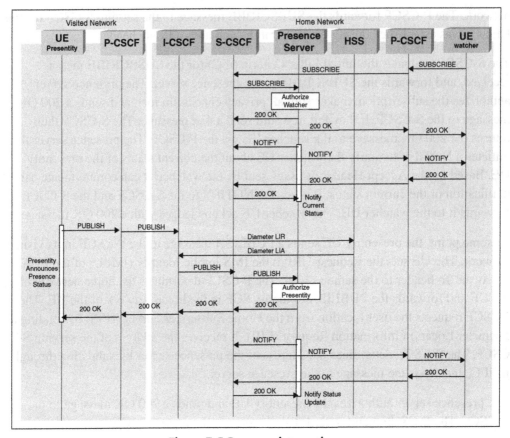

Figure 7.5 Presence interaction.

Agent (PEA), which can use the HTTP-based protocol to alter and obtain presence state information.

A PEA will use the following XCAP exchange with the resource list server to manipulate the resource lists, as shown in Figure 7.6.

7.2.2 Messaging

Even though IMS brings a rich set of voice and video means of communications, messaging still remains a vital paradigm. Messaging continues to be popular, as it can be carried out discretely and is less intrusive in terms of interaction instead of a voice/video

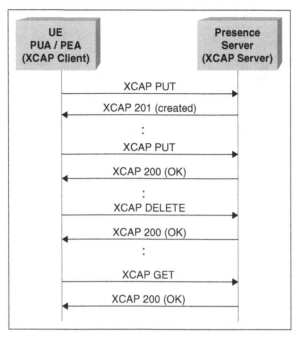

Figure 7.6 XCAP resource list manipulation.

call. Messaging has also allowed a scope for creativity by using short forms and codes. It has thus made its place as an indispensable communication method with the young crowd who use instant messaging and chat heavily. As this generation grows older, its reliance on a future platform such as IMS to support a strong messaging platform will weigh in.

Wireless service users have been exposed to chiefly two services: Short Messaging Service (SMS) and Multimedia Messaging Service (MMS). SMS is a point-to-point communication with messages getting delivered in quick turnaround. MMS delivers the message to the recipient's mailbox. It then notifies the recipient with an SMS message about the MMS message in his or her mailbox.

Internet users, on the other hand, use chat and instant messaging with presence as main forms of message communications. It is clear that IMS has to be able to provide methods to converge these various forms of messaging. The IMS network thus aims to leverage SIP-based methods to provide support for these paradigms. Three classes of messaging are established.

Figure 7.7 Enriched messaging experience.

- **Immediate Messaging** Delivery of messages is done in near real-time. It is also referred to as Page Mode, as it can allow broadcast/multicast of messages.

- **Deferred Delivery** The messages are held in store until the intended recipient becomes available. This mechanism is already provided by MMS. IMS leverages the MMS model for deferred delivery.

- **Session-based Messaging** The participants in a message exchange enter into a session such as a chat group or a conference.

Immediate messaging utilizes the SIP MESSAGE method and is useful to support instant messaging, inter-working with SMS, and so forth. Session-based messaging is useful to establish chat methods of communication and the ability to add audio, video, and document sharing in the session. While these methods form the basis of messaging, typically the user experience is enriched with presence and MMS methods. As shown in Figure 7.7, Sam checks for the presence of his friends who are available. He composes a multimedia message of his day at the beach and sends it to Don. Don receives the video message in his mailbox and initiates a chat session with Sam.

As we see in Figure 7.8, we consider a case with the message sender and receiver in two different networks. The originating UE initiates a SIP MESSAGE request to the P-CSCF. It sets the Request-URI to the IMPU of the intended receiver of the request. The To header is set to the same value. It sets the From header to its own SIP URI. The content-type header is set to plain text. The content contains the text message the user is sending.

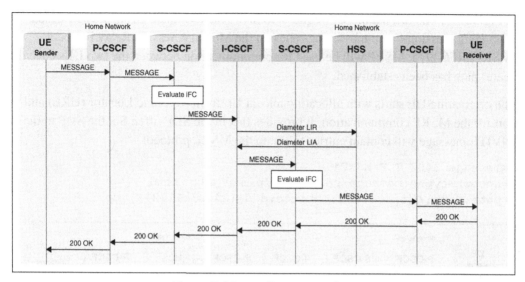

Figure 7.8 Immediate messaging.

The P-CSCF strips the necessary SigComp options, adds its route to the Route-Header, and sets the P-Asserted-Identity to the senders SIP URI, while removing the P-Preferred-Identity. It forwards the message to the S-CSCF.

The S-CSCF validates the subscriber profile and evaluates the initial Filter Criteria for the SIP MESSAGE. It determines the home domain network of the destination UE and sends the message to the I-CSCF. The I-CSCF sends a Diameter Location Info Request to the HSS to determine the serving S-CSCF for the MESSAGE. It then routes the message to the S-CSCF upon receiving the Diameter LIA from the HSS. The S-CSCF in the receiver's home network validates the subscriber profile and forwards the message to the P-CSCF with which the UE had registered. The P-CSCF forwards the message to the destination UE. Upon receiving the message, the UE sends a 200 OK message. This is then forwarded on the return path by the CSCFs to the originating UE.

A single immediate message can also be sent to multiple recipients. This can be done by including a list of URIs to identify multiple recipients. The URI list is passed in a multipart body in an XML format within the SIP MESSAGE request. The Request-URI in this case is set to the SIP URI of an application server. The AS can provide a resolution of the URI list.

To appreciate the differences between the immediate and session-based messaging, we consider a similar case of the sender and receiver UEs in different networks, as shown in Figure 7.9. The originating UE requests to invite the terminate UE in session-based messaging. Session-based messaging relies on the Message Session Relay Protocol (MSRP) RFC4975. MSRP will be used for communication between the two UEs once the bearer path has been established.

The originating UE starts with allocating a local Uniform Resource Locator (URL) and port for the MSRP communication. It provides this in the SDP offer. So, the SDP in the INVITE message will contain entries to set up the MSRP protocol.

```
m=message 3402 TCP/MSRP*
a=accept-types:message/cpim text/plain text/html
a=path:msrp://[5555::aaa:bbb:ccc:ddd]:3402/s111271;tcp
```

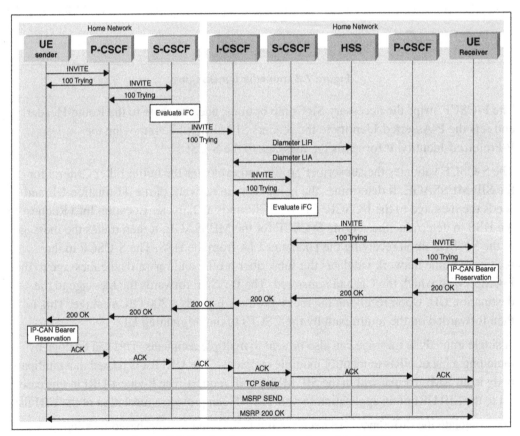

Figure 7.9 Session-based messaging with MSRP.

This, in effect, communicates to the destination UE about using MSRP on a TCP link on port 3402 in this case, and the capability to accept plain or Hypertext Markup Language (HTML) text message contents.

The routing and addressing of the INVITE is similar to the preceding example, to get to the destination UE. The entities at each hop return a provisional 100 Trying response to the previous hop entity. When the message is received by the receiver UE, it reserves the IP-CAN bearer for the messaging session media requested. It then confirms with a 200 OK message with the SDP answer acknowledging acceptance of the session and its readiness to listen on the TCP port requested by the originating UE.

When the 200 OK message is received by the originating UE, it initiates the IP-CAN bearer reservation, and sends an acknowledgement (ACK) toward the receiver UE. The originating UE continues to set up the TCP session on the IP-CAN bearer for the MSRP message exchange. Once the TCP communication is set up, it then sends an MSRP SEND to the receiver UE containing the text message. The receiver UE then acknowledges with an MSRP 200 OK message to indicate the receipt of the message.

This example illustrated the case where the MSRP session is directly established between the two UEs. It is possible to have the MSRP session set up with intermediate nodes. This could happen in situations where both UEs are roaming in different networks.

The ability to extend the session-based messaging into a conference or a chat room scenario is a promising feature over the immediate-messaging. This feature requires the UE to initiate a session to a conferencing server such as an Media Resource Function Controller (MRFC). The UE will send an INVITE request with the public service identity (PSI) of the MRFC. The S-CSCF will resolve the PSI to the MRFC according to its initial filter, and the criteria that will be handling the conference. Upon receiving the message, the MRFC will task the Media Resource Function Platform (MRFP) with an H.248 interaction to set up a conference bridge. Other users can join the session by initiating the session to the PSI of the MRFC. The Conference server also has the capability to invite participants to the session as instructed.

We now examine a scheme in Figure 7.10, which leverages the presence and messaging infrastructure we have discussed. This, however, goes beyond the realm of the standards. A converged messaging server functioning as a specialized application server can enrich both messaging paradigms with presence. It also extends the ability to inter-work with different

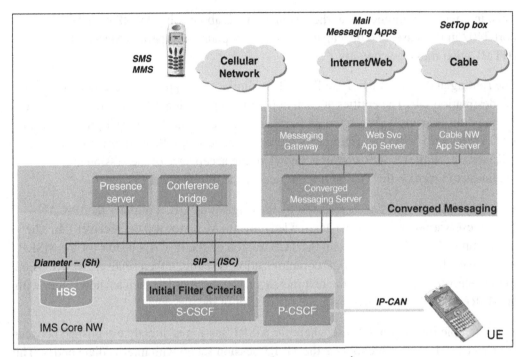

Figure 7.10 Converged message services.

messaging methods prevalent in other external networks. Let's consider a case where a message has to be delivered by a third-generation (3G) UE to a second-generation (2G) phone. Our previous scenarios showed us how to do this with two 3G UEs.

The 3G UE will need to initiate a SIP MESSAGE to the PSI of the converged messaging server, with the multipart body containing the identifier of the destination. The messaging server will interpret the destination identifier as a Tel URI, and request the services of a messaging gateway to inter-work the SIP MESSAGE to an SMS delivery to a 2G Mobile. A similar method would be used if the destination identifier is resolved as an Internet IM service address. The messaging server would then direct the request to a Web application server to deliver to the entity over the Internet services. This message approach would work for the reverse path as well. The message conferencing scenario explained can also be extended to external network users with this interworking model.

The Media formats and codecs for messaging and presence are defined in TS 126.141. Media elements can be expressed either as static or dynamic in nature. Multiple media elements can be combined into a composite single IMS message using MIME multipart content type format, as defined in RFC 2046 [25]. The media type of a single IMS message element can be identified by its appropriate MIME type, whereas the media format is indicated by its appropriate MIME subtype.

The following are forms of Static Media types.

- **Text** Expressed in plain text with character encoding from Unicode set.

- **Still Image** Support for ISO/IEC JPEG and TFIF formats.

- **Bitmap Graphics** GIF87, GIF89, and PNG formats.

The following are forms of Continuous Media types.

- **Speech** Adaptive Multirate (AMR) codec for narrowband speech. AMR-wideband codec can be used for support at 16KHz sampling frequency.

- **Audio** Support for one or both of the following codecs: Enhanced Advanced Audio Coding (AAC) and Extended AMR-wideband.

- **Video** ITU H.263 profile 0 level 45 is mandated. H.263 Profile 3, MPEG-4 Visual Service Profile and H.264 Baseline Profile Level 1b may also be supported.

7.2.3 Conferencing

Conferencing is the basis for multi party communication. It is also a fundamental component of enriching multimedia services by enabling group communication. While conferencing has been well adopted to collaborate in the enterprise environment in the CS networks, it has increased in popularity with the rise of social networking. Conferencing in IMS is defined in TS 124.147 standards. The recommendations, however, draw most principles from the IETF RFCs 4353, 4579, and 4582 for SIP conferencing. Most standardization work is being done in the Session Initiation Protocol INvestiGation working group (SIPPING) and the Centralized Conferencing Working Group (XCON).

As shown in Figure 7.11, the tightly coupled model for SIP conferencing is the recommended choice of the standards. In a tightly coupled model, the interaction between multiple participants is maintained by a central user agent called the "focus." This user agent identified by its own SIP URI, maintains a SIP dialog with each participant. The focus is responsible for ensuring the right media exchange between the participants.

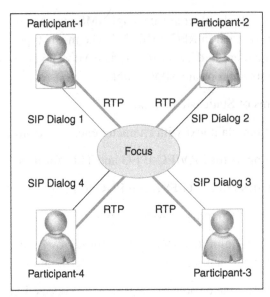

Figure 7.11 Tightly coupled model.

It also coordinates any conference policies to be applied. In contrast, a loosely coupled model does not have any coordination in the signaling relationship between the participants. Participation in the conference is done by multicasting arrangements. A fully distributed multiparty model can provide both signaling relationship and media between each conference participant.

We now examine the main components in the conferencing architecture, as depicted in Figure 7.12. Each conference has a unique coordinating user agent—the focus. The participants join the conference by sending an INVITE to the URI of the focus, and exit the conference by a BYE. The focus is responsible for maintaining these SIP dialogs with the participants. It also provides the capability to mix the real-time protocol (RTP) media streams. By applying a mixer, the focus can take a set of similar media streams, combine them according to specified rules, and then distribute to each participant. The set of rules that govern the mixing and in general the behavior of the conference is called the "conference policy." Conference policy is applied through a conference policy server. The policy server is a logical function that can store and apply the rules. It does not require a SIP interface. Participants subscribe to the conference through a notification service.

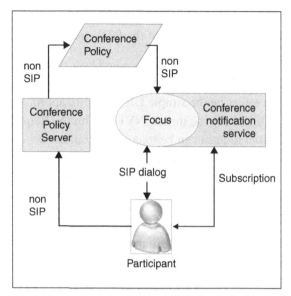

Figure 7.12 Conferencing architecture.

The Conference Factory URI is used to provide a method to instantiate a new conference and its focus. This is a globally routable URI, and establishing a call to this URI can automatically create the conference and the focus.

Once a conference has been established, it becomes essential to manage the access to the set of resources that are used. The resources are primarily the media sessions and components required to deliver the media streams. Control over these resources is provided by the floor control mechanism. The floor is provided by granting temporary rights to access or manipulate a set of resources in a shared or exclusive manner. Floor control provides the mechanism to obtain these rights in a safe and controlled manner. Access requests by clients to the floor are managed by the floor chair.

The physical realization of these logical concepts in IMS is as follows. The three primary components are the UE, the conferencing autonomous system (AS)/MRFC, and the MRFP. The UE provides the supporting functions for the conference participant. The conferencing AS/MRFC hosts the primary functions of the focus, conference policy, and media policy server. The MRFP provides the functions for the mixer and the floor control server. The conferencing AS provides the signaling coordination between the UE participants. It applies any policy and rules to control the access to the media resources. The MRFP provides the mixing capability of the RTP streams.

Let's examine the methods to set up a conference and how participants can join and leave a conference. A UE wishing to create a conference needs to provide the PSI of the conference AS. This can either be the simple URI or, more appropriately, the conference factory URI. The UE will address the initial INVITE with the Request-URI set to the appropriate PSI. This initial request can be to create a conference, or it can contain the list of participants to invite to the conference. In order to provide the list of participants, the Require header will contain the request list-invite, and the ContentType header will contain the URI list in the XML format. Media flows corresponding to the conference are provided in the SDP for the initial offer.

The IMS core network entities route the PSI and resolve it to the correct destination following the evaluation of the iFC processing at the S-CSCF. Upon receiving the INVITE for the conference setup and the initial SDP offer, the conferencing AS negotiates on the media flows and prepares to allocate the URI for the focus and the mixer. It sets up the H.248 session to control the resources at the MRFP. If provided with the list of the initial conference participants, it provides it to the MRFP to initiate the conference invitations to the participants.

Once the conference session has been established, a conference participant can join the bridge by establishing the session to the URI of the conference. The initiator of the conference can also now invite other participants. This is done by the SIP REFER method. The initiator will create a REFER message with the Request-URI set to the UE it wishes to invite. It provides the conference URI in the Refer-To header. It sets its own URI in the Via and the Referred-by headers. After the UE has acknowledged its acceptance, it sends a NOTIFY to the initiating UE about its progress and status about joining in the conference.

Exit from a conference is by a simple SIP BYE. The MRFP sends a NOTIFY to all the conference participants after it has released the resources allocated to the participant that has exited.

7.2.4 Push to Talk Over Cellular

Cellular Networks-based Push To Talk (PTT) is a half-duplex communication technique, which allows one-to-one or one-to-many group communication. The service is initiated

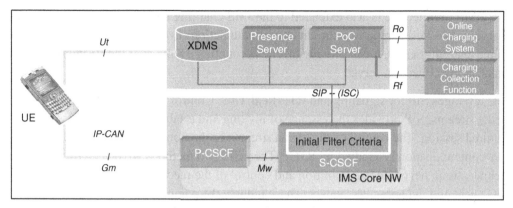

Figure 7.13 PTT elements in IMS.

by pressing a dedicated talk button without the need of dialing. The utility of PTT emerged from the land mobile radio networks. It became popular in industries with workers who are spread across different sites. The iDEN networks provided its foremost implementation in the cellular networks. Following the heels of its success, a consortium of companies defined a standard-based approach for the architecture for Push to talk over Cellular (PoC) networks. They delivered two Releases, 1.0 and 2.0. Open Mobile Alliance (OMA) adopted the standard, and PoC 2.0 sets the basis for PoC to use SIP enabled with IMS networks. While PTT has been implemented with non-SIP signaling, PoC 1.0 onward recommended SIP as the signaling model. This has helped PoC to integrate well within the IMS architecture.

Given the set of multimedia that IMS enables, PTT has morphed into Push to X services, X being a wildcard for talk, video, message, and even games. The model is extensible to invoke any of these multimedia services by pressing a single dedicated button.

The PTT architecture in IMS is defined in TS23.979, and is based on the OMA PoC architecture as described in OMA-AD-PoC-V2_0 (Figure 7.13). The PoC server extends itself to the IMS core network as an Application server, which communicates via a SIP ISC interface. The Presence server as we discussed in the previous sections is leveraged to provide the presence capability to the PoC Client. Group and list management as required for the PoC servers, are managed by the XML Document Management Server (XDMS), as defined in the OMA architecture. For the PoC user to access group and list information, it interfaces with the XDMS over the Ut interface.

The OMA architecture is also extended to the IMS charging methods of offline and online charging. The PoC can perform credit control interactions over the Ro interface to the online charging server. It can provide offline accounting events over the Rf interface to the charging collection server.

PoC leverages SIP for signaling and RTP/Real Time Control Protocol (RTCP) for the media streams. SIP provides the essence of inviting and establishing both parties in the required session. The control of the media streams or the floor control to provide the one-way communication is handled by RTCP. There are two methods advocated by the OMA architecture to negotiate the definition and control of the media resources between the PoC client and the home PoC server. These include the parameters such as IP address, ports, and codecs, which are used for sending the media and floor control packets.

- **On Demand** This requires a full session establishment and negotiation each time a PoC user initiates, establishes, or joins a PTT session.

- **Pre-established** This mode supports a pre-arranged agreement on the media resources prior to establishing the session. The PoC client can activate the media resources when needed.

Figure 7.14 is an example PoC session established on-demand with a manual answer. The scenario describes two UEs in their home networks. To simplify, we are abstracting the CSCF interactions as the IMS Core Network. To start the session, the UEs power on, perform a General Radio Packet Service (GPRS) session establishment, and register themselves with the IMS Core Network. During registration, the UE will register with a feature tag +g.poc. talkburst to indicate that the UE is capable of handling PoC session requests.

The initiating UE presses the talk button, which results in the PoC client on the UE to send a SIP INVITE to the intended destination UE. The Request-URI is set to the IMPU of the destination UE. In the case of a group session, it would be addressed to the PSI hosted by the PoC server. The Accept-Contact Header is set to +g.poc.talkburst, and the Required header is set to pref. Based on the two modes discussed earlier, the SDP will either contain a full description of the media and floor when this session is being established, or simply the IP address of the media, which was exchanged earlier.

When the message is received by the S-CSCF, the service indicator of PoC matches the iFC. The S-CSCF correspondingly routes it to the PoC server within the home network. The PoC server determines the home network of the destination and requests the INVITE

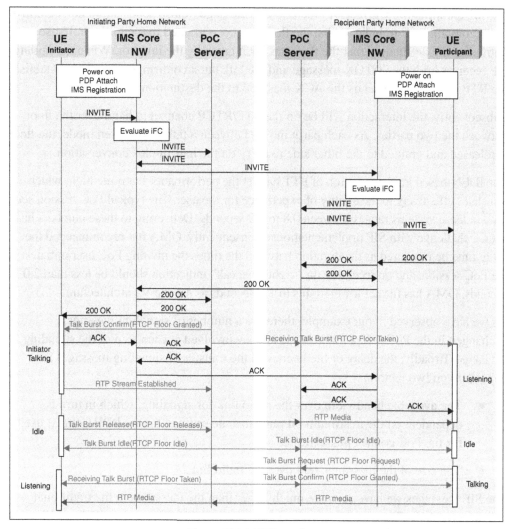

Figure 7.14 PTT session.

to be routed to the IMS core network of the recipient. The S-CSCF in the recipient's IMS network evaluates the filter criteria and routes the message to its PoC server. The PoC server then routes it to the destination UE. The UE, upon receiving the INVITE, makes a determination to the SDP offer if any adjustments are required for the media. It responds with a 200 OK message, which follows the similar return path. When the PoC server serving the initiating UE receives the 200 OK message, it also sends a talk burst confirm to the initiating UE to indicate that the floor has been granted to the initiator. It also notifies to the destination that the floor has been taken by the initiator. When the initiating UE receives both the 200 OK message and the talk burst confirm, it is ready to establish the RTP media flow. It sends the ACK message to the destination UE.

Subsequently, the interaction will be on the RTP/RTCP channel to hand over the floor between the two parties. As each party moves between a talk and a listen mode, the floor is released and granted to the other side to carry on the half-duplex conversation.

The iDEN-based implementation of PTT has set the performance bar quite high, which translates effectively to the quality of experience for the user. The typical PoC session setup time in these systems ranges between 0.8 to 1.2 seconds. Delivering to these numbers has been a challenge with SIP implementations. Consequently, OMA has recommended the setup time be measured as the duration between the times the inviting PoC user initiates the PoC session, and on receiving the "right-to-speak" indication should be less than 2.0 seconds. OMA has further lowered this to 1.8 seconds as per the V2 architecture.

As we have observed in our example, there are a number of signaling messages exchanged in the PoC setup, and there is latency involved at each step of the signaling exchange. Broadly, the delay or the latency in the transfer of signaling messages is dependent on two factors:

- The available bandwidth over the radio link for signaling, which in turn is dependent on the maximum bit rate attribute settings for the PDP Context used for the PoC control plane traffic.

- The size of the messages sent over the radio link.

The SIP messages we have seen are much larger than the messages in the traditional cellular protocols. The size of the typical messages used including the UDP and IP overhead is an INVITE of the order of 1400 bytes. A 200 OK message is of the order of 800 bytes. A 100 Trying is of the order of 300 bytes. The talk burst confirm is of the order of 40 bytes. Applying SigComp to reduce the size of these messages sent over the radio link is instrumental in reducing delays. The SigComp operation itself takes 50 ms to compress and decompress an operation. TS 23.979 calculates about 600 ms of delay permissible by SIP signaling, which translates to a requirement for 3:1 compression ratio to meet the 2-second goal.

7.2.5 Video Sharing

Video sharing is an application that enables users engaged in a voice conversation to turn on a unidirectional video stream and exchange it with the other party. The astute reader may wonder about its rationale. With IMS, it is possible to initiate a video call, so why do this in two steps? This paradigm actually harnesses an advantage by using a different implementation method. Video sharing allows the conversation to be carried out as a circuit-switched call and the video sharing as exchange of IMS signaling on the packet-switched network. Video sharing emerges as a forerunner IMS application that enables the subscribers in the CS-networks to get a taste of IMS. This also allows the service providers to monetize a new feature in IMS, and extend it to non-IMS users.

In a video call, the use of both audio and video for the call are established at setup time for the complete duration of the call. In a video share, however, the call involves adding or removing video sessions from a voice call. Video sharing provides the effect

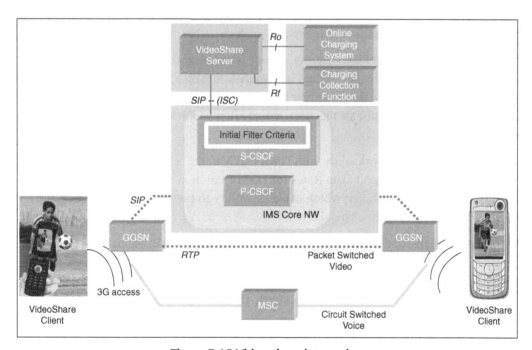

Figure 7.15 Video share interaction.

of enriching the conversation with visual means. A party engaged in conversation may choose to respond to a question by showing something on the video rather than just talking about it. A simple question like "How's the game going?" can be responded to by turning on the camera and pointing to the field (Figure 7.15).

The user experience with the video sharing is to either turn on the camera in the phone or play a recorded video stream and continue the conversation on the voice channel. When video sharing is turned on, the phone will automatically switch to speaker or headset mode, which allows the parties to talk while capturing or viewing the video.

Video sharing has been standardized by the GSM Association in its IR.74 standard, which allows building an interoperable service. Video sharing uses the concept of a combination of CS and IMS (CSI) services defined in TS 23.279, 23.279, and 24.279. A CSI session is a multimedia session that uses the CS domain to transport all or some media components, which is typically voice, and the IMS/PS domain to transport the other media components. A CSI session can be created either by first establishing a CS call and subsequently a concurrent IMS session(s), or by first establishing an IMS session(s) and subsequently a concurrent CS call. To the user, the experience of a CSI session is no different from a single multimedia session.

The implementation of the video sharing service is evolving from a simple user-to-user sharing of a unidirectional video stream in conjunction with a circuit-switched voice call to more advanced features. By extending the capabilities of the video share application server, it can support a multicast mechanism and allow video streaming to and from a Web or streaming server.

Video sharing currently works only with Universal Mobile Telecommunication System (UMTS) phones or Enhanced Data Rates for GSM Evolution (EDGE) devices with a dual-transfer mode capability, which is the capability to transfer both CS and PS signaling on the same radio channel.

We now explore the setup of a video sharing call between two parties, as shown in Figure 7.16. Technically, combinational services can be invoked in any order. The benefit of video sharing is, however, seen as adding video to an already established CS voice conversation, which is the flow we will examine. This comprises of the following steps:

- Establishing a CS voice conversation

- Capabilities exchange query

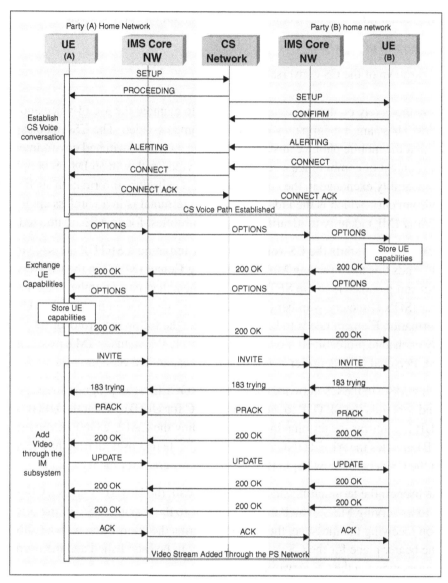

Figure 7.16 Video sharing.

- Invitation for a video session and negotiation on media resources

- Teardown of the video

- Teardown of the CS conversation

The CSI methods rely on using media feature tags to connote the use of voice and video capabilities. These are +g.3gpp.cs-voice and +g.3gpp.cs-video. The CS-video capability indicates that the mobile phone can support video in a circuit-switched environment within the context of combining a circuit-switched video call with an IM session. To enable a capability exchange in the context of the CS callsetup, information such as the radio environment – which is the indication that the terminal is in a wireless environment; Mobile Entity (ME) identity to identify known capabilities, for an ME are utilized as well.

The initiating UE(A) starts the CS voice session by initiating a SETUP message to UE(B). It sends this to its serving Mobile Switching Center (MSC) in the CS Network. The significant parameters in the SETUP include the called party number; i.e., the Mobile Subscriber ISDN Directory Number (MSISDN) of the destination mobile. The User-User Information Element needs to be set to indicate the following Protocol ID = 3GPP capability exchange protocol: Radio Environment = 1, IM Status = IM subsystem registered, Personal ME Identifier = 0007, and UE capability version = 04.

The Serving MSC in the CS network indicates a PROCEEDING progress message to UE(A) and forwards the SETUP to the serving MSC for UE(B). When the UE(B) receives the SETUP, it confirms the terminating call by sending the CALL CONFIRM message. The UE(B) initiates the ALERTING message to indicate that user alerting has begun. The MSCs in the CS network now begin to allocate the resources for the user plane.

When the user on the terminating side answers the call, the UE(B) initiates a CONNECT message to its serving MSC, which in turn routes it to the serving MSC of the originating side. Upon receiving the message, the UE(A) confirms the connection with a CONNECT ACK. The bearer plane for the CS call is now ready to be established and the users are in voice conversation via the CS network.

The capabilities exchange of the UEs to understand their capability to participate in a video sharing session can be done by a parameter exchange in the CS call. We discussed these parameters in the User-User Information element previously. The exchange can also be done outside the CS call, when the CS call is already in progress via the SIP OPTIONS method.

The originating UE(A) initiates a SIP OPTIONS message to UE(B) to know about its voice and video capabilities. It addresses the OPTIONS message to a Tel-URI destination. It also sets the Accept-Contact header with the media tags *, +g.3gpp.cs-voice, +g.3gpp. cs-video to inquire about the voice and video capabilities. The S-CSCF in the home IM network of UE(A) replaces the Tel URI to a SIP URI and forwards the message to the destination UE(B) home IMS Network. The UE(B) upon receiving the OPTIONS message stores the basic UE(A) capabilities and responds with a 200 OK message.

In the 200 OK message, it responds with its capabilities in the contact header and also the SDP. The UE(B) may indicate that it can support only CS-voice and recommend the supported video-codec. The UE(A) upon receiving this information stores it for further use. The UE(B) to obtain detailed capability information from UE(A) initiates a similar SIP OPTIONS message exchange.

At the next stage, when the UE(A) wishes to commence the video session, it follows the basis, as discussed in Figure 6.3, using the SDP offer/response initiated with the INVITE session. It creates the INVITE toward the Tel URI of the destination UE(B). It declares the session as a video session. It includes the "precondition" and "100rel" options tag in the supported header.

It provides its capabilities for CS-voice and CS-video, and requests to communicate with CS-voice or CS-video capabilities by including these media feature tags in the accept-contact header. It also includes its personal Mobile Entity (ME) identifier and UE capability version in the user-agent header. It provides the initial set of video and audio capabilities in the initial SDP offer.

Subsequently, the two UEs negotiate on the supported media flows in the SDP offer and response in the PRACK and UPDATE message exchange. Upon acceptance of the necessary media flows, the UE(B) responds with a 200 OK message for the initial INVITE and the UE(A) sends an ACK. The Video stream is then ready to be exchanged.

Should the UE roam from a 3G area to a Global System for Mobile (GSM) area, the video session will be torn-down but the CS-voice will continue. A graceful shutdown of the call will first bring down the video session and then the CS-voice session.

7.2.6 Voice Call Continuity

Fixed Mobile Convergence (FMC) has evolved as a de jure standards approach focused on enabling traditional wireline voice networks to converge with wireless and mobility features.

During the course of its evolution, the scope of wireline voice expanded to include broadband IP data and Voice over Internet Protocol (VoIP) including cable network access. Subsequently, FMC surfaced the opportunity for an application, which could allow a single device switch service seamlessly between a Voice over wireline/IP/Cable network and the wireless cellular network. 3GPP proposed the concept of Voice Call Continuity (VCC) in Release 7, to provide an effective solution and to apply it in the IMS context (Figure 7.17). 3GPP2, TISPAN, and PacketCable 2.0 standards extended their support to this application as well.

VCC is a residential IMS application that allows a UE to switch between an IMS and a circuit-switched domain, while providing a seamless experience to the user while on a voice call. The user experiences continuity in a voice call regardless of the transition in the underlying access network. VCC defines the functions for call origination, termination, and the domain transfers between IMS and the CS domain.

VCC is accomplished by the concept of *anchoring*. Anchoring is the process of creating, joining, and maintaining two connections for the purpose of service continuity. The anchoring is provided by a VCC application server in the home IMS network. Voice calls to and from a VCC-capable UE while in the CS domain or in the IMS domain are anchored at the VCC application server. The VCC application server applies third-party call control to enable an inter-domain transfer. The VCC application server can perform an inter-domain transfer multiple times within the same call session.

The VCC application server consists of four main functions. The Domain Transfer Function (DTF) executes the necessary logic, with third-party call control to execute the transfer of the UE access between IMS and the CS domain as requested by the VCC-enabled UE. The Domain Selection Function (DSF) is invoked to determine the domain to be used for a VCC-enabled UEs incoming call. This selection is determined on the basis of the operator policy or subscriber preference. The CS Adaption Function (CSAF) performs the role of a proxy to maintain CS originations, CS legs, and domain transfers. Finally, the CAMEL Service is invoked for the CS call originations or terminations in a roaming network. The VMSC requests the address information from the gsmSCF function to route the call to the home network.

The VCC-capable UE has the capability to support voice over a CS domain and IMS domain. The UE is responsible for obtaining and storing the user preferences and operator policy. It can then apply these for domain selection for a call origination and for a domain transfer. The operator policy is given precedence over the user preferences.

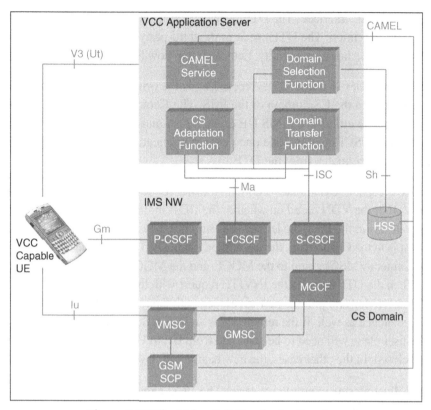

Figure 7.17 The Voice Call Continuity function.

VCC introduces four new identities, which are required to support the routing and transfer function. The CS Domain Routing Number (CSRN) is the identifier used for routing a call in the CS domain to the user in the CS domain. The format is similar to the MS routing number (MSRN). The IP Multimedia Routing Number (IMRN) is used to route the call to the IMS network. It can be expressed as an MSISDN or as a Tel-URI. It is handled as a PSI within the IMS network. The VCC Domain Transfer Number (VDN) is a E.164 number used by the UE to request a transfer to the CS domain. The VCC Domain Transfer Identifier (VDI) is a Tel URI used in a similar manner by the UE to request a transfer to the IMS.

The control to the functions in the VCC Application Server is executed by the S-CSCF iFC service invocation mechanism. For a CS-originating call, control is given to the CS-Adaption

Function over the ISC interface. The DTF is invoked as a part of the originating or the terminating iFC execution. The DTF executes the third-party call control function, which provides the static anchoring capability. Let's examine how this works.

A VCC call comprises two legs. The Access Leg is the originating leg, which can originate either from the CS domain or the IMS. The Remote Leg is the terminating leg, which can terminate either to an IMS UE or a CS terminal. The DTF maintains both the legs. While the S-CSCF may include one or more application servers for the remote leg, its signaling path originates from the DTF.

When the VCC capable UE initiates a domain transfer request, it sets up a new access leg. It uses the VDI or the VDN based on whether the UE is in the CS-domain or IMS domain, respectively. If it is in the IMS, the leg will be established by sending an INVITE to the S-CSCF. If it is in the CS-domain, the leg will be established through the Visited Mobile Switching Center (VMSC) going to the MGCF, and the MGCF will send the INVITE to the S-CSCF. When the DTF receives the INVITE request with the new access leg, it will release the original access leg and join the new access leg with the remote leg. The bearer plane needs to be switched as well. If the remote leg is an IMS UE and the VCC UE has transferred to IMS, the user plane will need to be switched from UE to UE and the connection through the MGCF closed. In the other cases, the new bearer will have to be connected at the MGCF.

Figure 7.18 illustrates an example scenario. The VCC capable UE establishes a CS Access by connecting to the VMSC in the visited CS network. The VMSC initiates an origination request to the gsmSCF, which requests the routing authorization from the home IMS network. The VMSC directs the call signaling to the MGCF, which translates the ISUP CS signaling to SIP and directs it to the S-CSCF. The S-CSCF invokes the VCC application server to anchor this call. Since it is a CS origination, the call is actually routed to the CS adaptation function and then handed over to the DTF for maintaining control. The DTF initiates the signaling for the outbound leg, which in our example is an IMS endpoint. The S-CSCF in this case includes another AS in the remote leg establishment. As illustrated, the bearer goes through the CS Network and through the Media GateWay (MGW) to the IMS endpoint. When the VCC capable UE wishes to initiate a domain transfer, it now sets up an IMS access leg identified with the VDN. This leg is set up and maintained by the DTF. Using third-party call control, the DTF now switches the CS access leg to the IMS access leg. The bearer in this case is straightforward, which is set up between two UEs as they are in the same IMS domain.

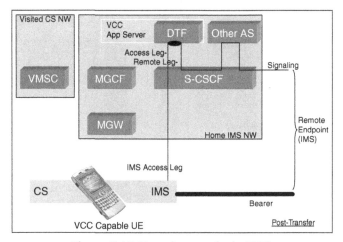

Figure 7.18 Domain transfer in VCC.

7.3 Enterprise Services

Private Branch Exchange (PBX) or the hosted Centrex systems (CPE) have been the backbone for communication systems in the enterprise for several years. While the technology behind these has evolved from analog to ISDN to IP, their functionality has remained restricted to voice only services. Enterprises, on the other hand, are critically dependent today on e-mail and other collaboration methods. Mobility has also become increasingly important to different segments in the enterprise. The challenge in the enterprise is the disparate methods for delivering services. IMS provides a common platform to enable a unified set of collaborative services in the enterprise. While the basic set of presence, messaging, and conferencing is equally applicable to the enterprise, we examine the services that bring a rich user experience in workspace collaboration.

We examine a model in Figure 7.19, for supporting enterprise IMS applications. A large enterprise normally comprises a central or a corporate office and one or more branch offices. The central enterprise hosts an S-CSCF, media functions, and the enterprise application servers. This is an adjunct to an Mobile Network Operator's (MNO's) main IMS network, and appears as an independent network. Each branch office is equipped with a P-CSCF. The interaction between an enterprise user and a mobile network user would be as follows. The enterprise UE discovers the P-CSCF in the branch office with the help of a DHCP server. The branch office P-CSCF enables the authentication and

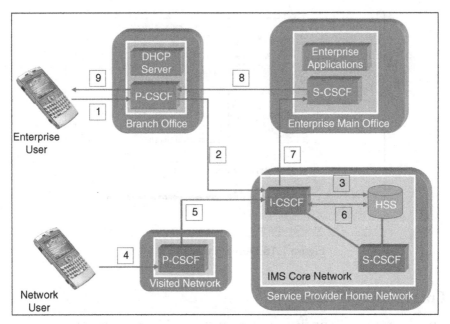

Figure 7.19 IMS in the enterprise.

registration with the MNOs home network. When a network UE wishes to communicate/
collaborate with the enterprise UE, it initiates a session to the P-CSCF of its visited/
home network. The I-CSCF then determines the location of the S-CSCF, which is in the
enterprise main office. The S-CSCF then routes the session to the registered enterprise
UE in the branch office. The enterprise UE can then invoke an application hosted in the
enterprise application servers. We now look at some of these applications.

7.3.1 Collaboration/Group Lists

Collaboration in an enterprise is enhanced with effective group management.
As explained in TS 22.250, group-related services are presence, messaging, and
conferencing. These are fast becoming indispensable enterprise communication tools.
Conceptually a group refers to a collection of users within a certain context sharing
certain properties. Group members can share their availability, contact information, and
ability to communicate via a selected method. The example illustrated in Figure 7.20
shows how users within a group can use different media forms for communications.
These forms could be voice, video, text messaging, e-mail, or IM. Group management

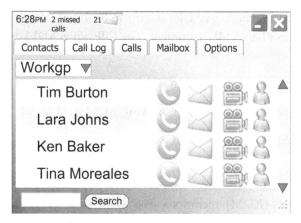

Figure 7.20 Unified phone book.

allows a single user interface for all communication channels. This obviates the need for multiple contact lists and reduces the overheads of duplication.

Group management is a simple yet powerful concept with the following features. A group is maintained by a specially designated user—a group administrator who has the complete set of privileges to maintain the group list. The members of the group are assigned a set of rights to access the capabilities of that group. Each group has a unique identifier and a set of attributes. The attributes are the information about the group and the properties of the group. Group visibility, which defines the member privileges and the group duration, is such an attribute.

Group membership requires an identity to identify the members within the group. Each member is granted a set of rights, which are the actions the members can perform. Anonymity is a right as well. The group management functions are the basic operations to create a group, delete a group, add members, get the list of members, modify the group service information, and perform a search on the group. Access to the groups is possible from multiple terminals.

As we observed earlier, access to group information in the presence and PoC models was enabled via the XDMS. The XDMS stores and manages the group membership information in an extensible XML data model. With the use of simple SIP and XCAP interfaces, the group information can be easily accessed by the application servers.

With the group lists managed in the XDMS, XDM clients can access and manipulate the group lists using the XCAP interface. XCAP allows clients to manipulate documents and to update documents using an XCAP URI (Uniform Resource Identifier) that references elements and attributes of an XML document. The following XDM operations are provided:

- Creating, replacing, deleting, or retrieving a document

- Creating, replacing, deleting, or retrieving an XML element

- Creating an XML attribute for an XML element, deleting or retrieving an XML attribute

The HTTP GET, PUT, and DELETE methods are used to perform the preceding XML Document management (XDM) operations. The XDMS also supports XCAP server capabilities and XML documents directory queries from XDM clients. Each document stored in the XDMS is associated with an Entity Tag (ETag). The ETag enables clients to determine whether the document kept in its cache is the most current. Conditional HTTP operations can be used to compare the ETags in the clients and XDMS prior to retrieving the document.

Figure 7.21 further illustrates the OMA- based XDMS architecture. The architecture defines an XDMS for specific services or enablers, and a shared XDMS to store common

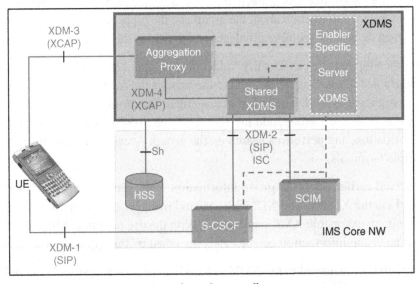

Figure 7.21 XDMS-based group list management.

data between the different services/enablers. In addition to the XMDS, it identifies an Aggregation Proxy, which is responsible for authenticating the XDM client and then routing to the appropriate XDMS. These XDMS are distinct from application servers. The interaction between these XDMS is done through the following interfaces.

The XDM3 and XDM4 are XCAP interfaces. The XDM3 interface provides the operation for manipulation of the XML documents to an XDM client. The XDM client communicates with the aggregation proxy to direct the requests to the correct XDM server. The XDM4 interface provides the operations for the manipulation of the XML data to the IMS core network entities. It allows access to the XML data in the shared XDMS.

The XDM1 and XDM2 are SIP interfaces. The XDM1 permits the subscription to document changes. Traffic will flow on the XDM1 only after it has passed authentication through the IMS CN. It interfaces to the application servers. The XDM2 interface provides similar functions between the IMS core network and the shared XDMS.

In addition, the XDMS can query the user profile from the HSS via the Sh interface. The XDMS also can interface with the charging system for billing, and support a provisioning interface.

7.3.2 Whiteboard, Application, and Document Sharing

Whiteboard, application, and document sharing are visual information exchange methods that enrich conference calling and collaboration in the enterprise. While Cisco-Webex and Microsoft LiveMeeting are effective means to provide collaboration via the Internet, IMS can provide a better platform enabling wireless mobility. Historically the ITU T.120 standards and a combination of protocols such as H.323 have been used for these conferencing methods. However, their complexity has limited their adoption. IMS does not have any standards dictating how to perform these applications, and the solutions are vendor-specific. It is interesting, however, to explore how this can be enabled by IMS.

The three methods of sharing employ a common principle. The view from an application or a document is captured as an image and transferred into a conference bridge for distribution to other participants. Whiteboard sharing, as shown in Figure 7.22, requires a multi-planar view, where a base image plane is imposed with an annotation plane for marking and a virtual plane for directing the cursor movements. A sharing client on the UE provides the capability to allow the user to select the application method and have the necessary bitmap image ready for exchange.

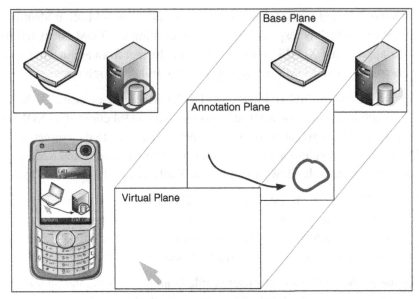

Figure 7.22 Whiteboard sharing.

This is where things become interesting. Transmitting still images is different from a video stream over an RTP media flow. Second, an interaction, if required, with the application server necessitates a higher level user-plane protocol. We will examine how MSRP can be effective for this communication. These collaboration applications can be implemented as CSI applications; however, it is not essential. We examine a collaboration scenario, as illustrated in Figure 7.23. For simplicity, we assume a single network with two UEs participating in this exchange. For a collaboration application, we require a collaboration client and a collaboration server. The collaboration server is a SIP A/S, which controls an MRFP. The communication between the SIP A/S and the MRFP can employ an XML-based protocol such as VoiceXML, supported through HTTP or XCAP exchange. This protocol is also effective for communicating any MSRP information received, for the UE collaboration client to exchange any user-plane information with the server.

UE(A) wishes to establish a whiteboard sharing session with UE(B). It initiates a SIP INVITE with the Request-URI containing the PSI of the collaboration server (A/S). The Require header will contain the request-list-invite, and the ContentType header will contain the URI for UE(B) in the XML format. The initial SDP offer in the INVITE contains the Content-Type set to image/jpeg and the URL for the TCP port on which

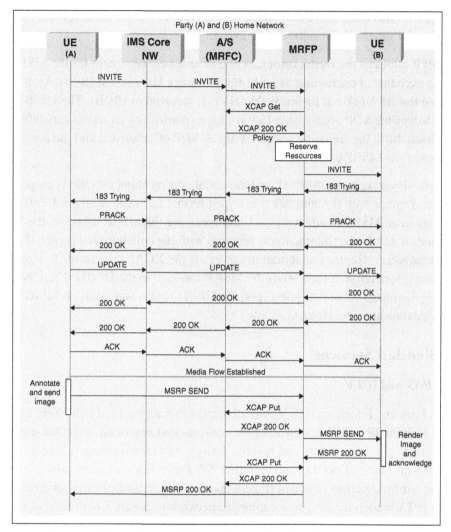

Figure 7.23 Collaboration applications flow.

UE(A) is willing to receive MSRP messages. When the INVITE is processed by the iFC of the S-CSCF in the core network, it routes the message to the intended collaboration server.

The collaboration server then directs the INVITE to the MRFP selected for handling this application. It provides the UE(B) address in the Request-URI, and the URL for the XCAP/HTTP message exchange. As mentioned, VoiceXML commands can be exchanged on this interface to communicate with the MRFP for the resources and description of the media permitted to be exchanged.

The MRFP allocates the media resources and establishes the stream for the MSRP message exchange. Functioning as a SIP Back to Back User Agent (B2BUA), it creates a new leg toward UE(B). It forwards the INVITE received to UE(B). The UE(B) acknowledges the SDP offer with a 183 Trying response with its media capabilities. The two UEs establish the signaling session with the SDP offer/answer and the ports for the exchange of the MSRP exchange.

The media flows are re-established, and the collaboration client on UE(A) prepares the whiteboard image with the annotation in a jpeg format for transmitting to UE(B). It sends this image in an MSRP Send message. Upon receiving the MSRP message, the MRFP determines if any further interaction is required with the collaboration server. If so, it can communicate with the collaboration server via the XCAP commands to provide this user-plane information. It then sends the MSRP message to the UE(B). The UE(B) upon receiving the image content, renders it on the UE(B) screen and sends an MSRP 200 OK acknowledgement to the User(A).

7.4 Blended Services

7.4.1 IMS and IPTV

Internet Protocol Television (IPTV) broadly refers to the system of delivering television content with the IP protocol, instead of the conventional broadcast and cable medium. IPTV aims to provide a secure and reliable delivery of video services. These include live TV, Video-on-Demand (VOD), and recorded TV. These video services utilize an access transport agnostic packet-switched IP network to transport audio, video, and control signals. IPTV is architected over a controlled network to ensure a high level of Quality of Service (QoS) and Quality of Experience (QoE). In other words, IPTV is not just video over specific technologies such as DSL or video streaming over the Internet.

The telecom service providers have found IPTV of competitive advantage in the video space. It has been offered as a bundled service referred to as a "Triple-Play." The transport agnostic nature of IPTV allows it to be delivered over a broadband connection such as DSL, or it can be delivered to a mobile device via GPRS or Evolution-Data Optmized (EV-DO). So where does IMS fit in?

Both IPTV and IMS share a common thread. They started with existing best-of-breed technologies and services. Then they laid these out on an IP network, which is agnostic of access transports. When you compare on surface there is nothing radically different from the way things are already done. What is different is that this now provides a flexible and more adaptable framework, which can easily be merged to provide innovative combinations. This is a key foundation of building a converged service.

While the current service models have focused on the technology-focused services, IMS and IPTV shift this to more personalized services. The current telecommunication service paradigm also steers from the bundling of services to blend them instead and provide seamless composite services. It becomes intuitive to see why IPTV differentiates over bundled services of Triple Play and promises services beyond Cable TV VOD.

The consumer is already used to live TV, channel services, subscribed channels, Pay-per-view, and VOD. In addition, the schemes for timeshifting through DVRs and TiVO and placeshifting such as using the slingbox, where a broadband signal can be retransmitted to a different location via IP, are well used. What the consumers have yet to see are more combinational services that blend in communication services with TV programming. Such services are:

- Interactive TV and SMS Televoting

- Web browsing

- Notifications

- Video calling

IMS complements IPTV to provide a communication platform and an access to a converged network. The IP fabric and transport independence allow the IPTV infrastructure to appear as a UE. With this service identity established, the IPTV can now get an access to converged networks that IMS enables. New IPTV applications for service invocation, delivery, and management are still required. That is still exciting as it opens new avenues that have been limited with the current generation of wireless and wireline networks.

Let's examine some possible services that can be enabled with an IPTV-IMS combination (Figure 7.24).

Figure 7.24 Blended IMS IPTV applications.

- **Interactive Programming** The popularity of *American Idol*, *America's Got Talent*, and similar shows around the world has primarily been because of the viewer's vote. For now, the viewer has to either SMS or call with a phone device. Calling at these congested hours results in waiting. An IPTV-IMS combination first eliminates the need for a separate device. The IPTV viewer is registered as a UE endpoint. The viewer can now use the remote to send the text message to vote. This brings a new level of participation to the TV viewer. Further, the IMS-based application server can load balance between SMS or SIP message interfaces, to ensure the peak loads do not cause congestion.

- **Interactive Shopping** For commercial programming such as the Home Shopping Network or QVC, the viewer can use the TV as a calling device instead of a phone line. The same applies to local advertising for ordering pizza. This provides more convenience and ease of use.

- **Info Streaming** During a commercial break, Mary wants to check if her son Chris has left for home after his basketball practice. She sends an invocation for the locate service from her remote. An IMS application server obtains the location coordinates of Chris, plots it on a map, and streams this as Picture-in-Picture on the TV screen.

The standardization of IMS and IPTV is in the initial stages at the time of writing. The ETS-TISPAN TS 182.027 defines the protocols and interaction for IPTV support by IMS. We examine a generic model based on this foundation, as depicted in Figure 7.25. Since

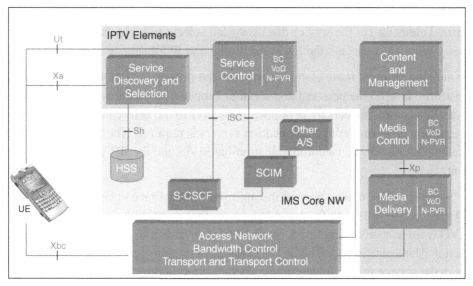

Figure 7.25 The IPTV-IMS model.

both IPTV and IMS are access network agnostic, the model can work with the supporting IP-CAN for the transport and bandwidth control.

The IPTV framework comprises a set of elements to support a set of three basic services:

- Broadcast (BC), which is the support for live TV services

- VOD or Content on Demand (COD), which allows the capability to access and view content on demand by the user

- Network Personal Video Recorder (N-PVR), which allows for storing content and viewing at a later point in time

Each of these services can be logically partitioned into three functions: Service Control, Media Control, and Media Delivery. The Service Control Functions interface to the IMS network at the application services layer. This function provides the IPTV service selection and authorization including credit control. It is responsible for selecting the right IPTV media functions. The IPTV Media Control Function provides the control of the media flows and the media elements responsible for processing them. The media delivery function is responsible for the storage, processing, and delivery of the media flow.

The European Telecommunications Standards Institute (ETSI) model identifies specialized elements for the service discovery and selection. It also provides the necessary functions to access the IMS user profile and IPTV service profile.

While the interaction of these elements provides the delivery of IPTV through the IMS as a platform, how do the personalized services happen? Given the architecture where the IPTV service control appears as an application server to the IMS core network, creating the interaction between multiple application servers is required. The SCIM provides the service interaction between a communication-based AS and the IPTV service control element.

As an example, let's examine how an incoming call indication can be provided during a VoD session. We follow on from a video stream that is currently in progress. When an INVITE for a session is received, the S-CSCF forwards it to the Service Capability Interaction Manager (SCIM). The SCIM determines that this requires service interaction. It checks the service profile to determine whether the called party has selected the option to receive an incoming call indication during the VoD, and is not set to a do not disturb. It then indicates to the IPTV service control server to halt the VoD session momentarily. It forwards the INVITE to the voice application server. Lastly, the SCIM then coordinates with the MRFC to stream the image for the caller to the user.

7.4.2 Location-Based

The precise situation of the user in terms of the geographical position information can enrich the blended services IMS can offer. While the IMS architecture provides the storage of the location information in the HSS, it has to utilize its application plane to gain access to location information. There are no specialized elements in the IMS architecture for location services. The location architecture is provided by the 3GPP/3GPP2 and OMA standards, respectively, for network-based and handset-based Assisted Global Positioning System (A-GPS) solutions, respectively. IMS therefore needs to inter-work with these architectures to obtain the positioning information. The standardization for this integration is still in process and we will discuss some approaches.

An IMS UE can either be a mobile entity or a fixed element. The location of a fixed element is relatively easier to determine by looking it up in a configured address database. VoIP networks determine the location for Public Service Access Points (PSAPs) on this principle. Coarse location of a mobile can be determined by its cell-id. The high-accuracy position of the mobile UE in terms of its latitude and longitude coordinates and velocity requires sophisticated methods to triangulate the precise location.

Network-based technologies use the methods of Timing Advance, Angle of Arrival (AOA), and Uplink Time Delay of Arrival (UTDOA), to name a few. The Location Measurement Units (LMU) in a 3GPP network or the Position Determining Entity (PDE) in the 3GPP2 network obtain these Radio Frequency (RF) measurements by a passive overlay network to the Radio Access Network (RAN). The algorithms applied result in close approximation of the location.

Handset-based technologies utilize the Generalized Processing Sharing (GPS) infrastructure. The UE must be enabled with GPS capabilities, which it can provide to the location servers. The OMA Secure User Plane (SUPL) enables the position determination assisted by the handset. The SUPL Positioning Center (SPC) provides the computation of the location to service a request received either from the handset or the network via the SUPL Location Center (SLC). GPS and A-GPS methods have limitations of indoor coverage. A hybrid solution with both network and handset approaches in place are being implemented as well.

These methods do become transparent to a requesting entity, which in our case is the IMS application. A QoS parameter can be provided, which controls the precision of the location returned. The selection of the method to obtain it is, however, not within control.

At a high level, the two main tasks are how to present IMS as a requesting entity to the location network and how to obtain location information. This requires supporting the interface to the location network, and presenting the identity of the UE in the form of a Mobile Station Identity (MSID). Following is how to disseminate the location information once it is available. The SIP extensions in IMS do not include support for location and privacy, although the Geopriv headers in SIP are under implementation.

The location services architecture provides a client-server approach where a client requesting location information can request the location server (Figure 7.26). The Gateway Mobile Location Center (GMLC) in a 3GPP Location Service (LCS) network, the Mobile Positioning Center (MPC) in the 3GPP2 Location Based Services System (LBSS) network, and the Secure user plane Location Center (SLC) provide the service to an external location service client. While the technology used in these location networks may be different, the approach to service a client request is similar. Most implementations of the GMLC, SLC, and MPC do support a common protocol implementation which is the OMA-LIF Mobile Location Protocol (MLP).

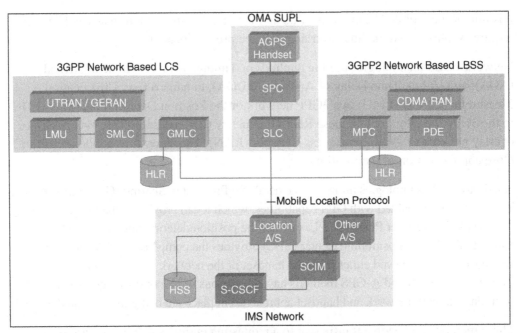

Figure 7.26 Obtaining location information.

MLP is an XML-based protocol that allows a set of services for a location request. These are immediate delivery, deferred delivery, and emergency services. Immediate delivery returns a location report in the response to the request. Deferred delivery or a triggered request allows a mechanism to track the user by registering for receiving one or more location reports upon the expiry of a time interval or a condition such as the mobile becoming available. Emergency location provides a higher priority location service for immediate delivery.

We now examine the sequence of steps to obtain the location information, as illustrated in Figure 7.27. To start, we assume that the UE has appeared in an RAN access network. It has powered on, established a data session, and registered with the IMS network. As part of a service invocation that requires location information, the location application server would need to request the position of the UE from the location network. The location AS can receive a request from another application server, brokered by the SCIM. The location AS first queries the HSS with a Diameter User Data Request. The informed reader will recall the User Data, defined in Section 5.3.3. The Sh user does contain the position information for the UE registered either in the CS or the processor sharing (PS) network. While the TS 28.329 mentions that the HSS can obtain this information

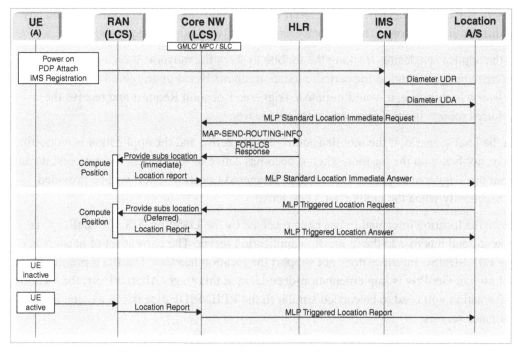

Figure 7.27 Flow for obtaining location information.

by a MAP-SUBSCRIBER-REQUEST from the Home Location Register (HLR), this implementation is not mandated in the current HSS implementations. The location AS can determine whether this information is available and is current. Otherwise, the AS determines the MSISDN corresponding to the IMPI that it requested the user data.

It then creates an MLP request to obtain the location. In our first scenario, the AS requests for an immediate location. It therefore sends a Standard Location Information Request to the location server in the core network, corresponding to the access RAN or technology used. The location server would check the privacy settings for the user, if location is permitted. The location server, the GMLC for instance, would need to determine the address of the serving MSC/SGSN for the UE. It would therefore request the HLR with a Send-Routing-Info-For-LCS to obtain the address. It then directs a message Perform Subscriber Location to the MSC. The MSC then sends an invocation to the Serving Mobile Location Center (SMLC) or the Position Determining Entity (PDE) cluster with a request. In 3GPP2, the BSSAP-LE protocol is used for this purpose. The SMLC, PDEs, or the SPC corresponding to the location network compute the position based on the QoS selected. The Location Report would contain the coordinates and any additional information such as velocity or timing advance that is requested.

If the application desires tracking the mobile to show the movement on a map for navigation, it can request for periodic location reports based on triggered or deferred delivery. In this case, it would initiate a Triggered Location Request and receive the deferred reports in the Triggered Location Answer.

In the final scenario, if the mobile is currently not active and the application is requesting to be notified with the location when it becomes active or registers in the network, it can send the Triggered Location request. The Triggered Location report will be provided subsequently when the mobile becomes active.

Once the location information has been received by the Location A/S, the challenge is how to send it across to the requesting application server. The current set of headers in the IMS SIP ISC interface does not support the location headers. The IETF proposed solution of GeoPriv is implementation-dependent at this stage. Alternatively, the location information will need to be carried similar to the RPID/PIDF extensions as seen in the presence servers.

7.4.3 Monitoring

We now examine blending a streaming application with IMS. Video surveillance systems are ubiquitous in industrial, residential, retail, and airport facilities to provide a secure environment. Most analog solutions have evolved to digital systems. The video from these systems is transmitted to a control center for monitoring. For a mobile user community, including security personnel the ability to get this video stream on a handheld device makes it easier for staff to monitor while moving about. Other novel applications of being able to monitor one's own residence or a child-care facility by simply calling from a cell-phone, enhances a simple surveillance system.

Video surveillance systems, as depicted in Figure 7.28, support a variety of cameras and interfacing methods. Typically these are sophisticated close-circuit TV (CCTV), IP cameras, or Web cams connected via Ethernet , USB, or 802.11 wireless. The video acquisition server or a video grabber provides the capability to obtain the video inputs from one or more of these camera types. The video acquisition server preprocesses the video feed encode and compress and has the stream ready for transmission. Prevalent codecs such as H.263, H.264, MPEG-2, and MPEG4 help in encoding the acquired video inputs.

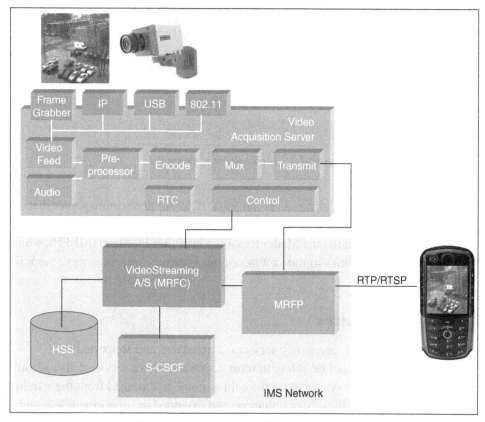

Figure 7.28 Video surveillance with IMS.

Preprocessing the captured video prior to encoding by low-pass filtering, can significantly reduce the noise level prevalent in industrial and public environments. Post-processing is used to reduce the artifacts generated during video coding. This step can improve the video quality significantly in a low-bit-rate transmission case. Feed from a real-time clock RTC also provides the timestamping on the video stream.

The multiplexer module combines multiple channels originating from different cameras or different angles. The stream is then ready to be transmitted on an IP transport (Figure 7.29).

To obtain access to this stream by an IMS UE, three functional elements play an important role. The UE must be able to host a client that can support a streaming control protocol. The real-time streaming control protocol (RTSP) provides the control for the

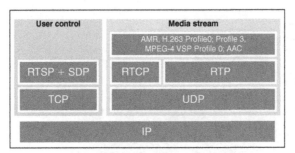

Figure 7.29 Media stream elements.

media stream, which will be delivered via the usual RTP. An application server that also functions as an MRFC and provides the control to the video acquisition server is the second component. Finally, the Media Resource Function Platform (MRFP), which provides the media capabilities to mix, transcode, and deliver these streams to the RTP stream to the UE.

7.5 Emergency Services

The requirement to support emergency services is mandated and supported by government legislation for public safety in most countries. These services, also identified as 911 (in US) and 112/999 (Europe) by the calling number, emerged from the wireline networks. Since then these have been enhanced and extended to support wireless and VoIP networks, and also referred to as E911/E112. The concept to support emergency services comprises two steps. First, any user must be able to get access to the network when the emergency number is dialed. Second, the call must be routed with its identity information to a public safety access point (PSAP), where the call can be serviced.

This concept works well with a fixed access or a wireline network. The address and location of each endpoint is fixed. The switching infrastructure can therefore route the call to the PSAP configured to support endpoints from a particular access area. Once the call reaches the PSAP, it looks up the Automatic Number Identification (ANI) in an Automatic Location Information (ALI) database and its companion Master Street Address Guide (MSAG). It can thus obtain the exact street address of the phone where the call was made.

Mobility with a wireless network poses a twofold challenge. First, which PSAP should the call be routed to that serves the current area where the mobile user is present. Second,

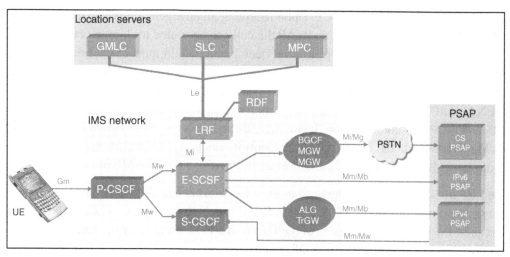

Figure 7.30 Emergency services architecture in IMS.

obtaining the accurate location of the mobile as there is no ALI database that can be used in this case. Wireless 911 and VoIP networks have working solutions for these.

IMS has been able to address support for emergency services only from Release 7 onward. Emergency sessions are described in TS 23.167. The supporting architecture, as shown in Figure 7.30, for emergency services identifies two new elements. These are the emergency CSCF (E-SCSF) to handle the emergency sessions and the Location Retrieval Function (LRF).

The UE conducts some vital functions while initiating the emergency session with the IMS network. The UE can detect the request for an emergency call. It needs to allocate a special public user identifier for the session such as anonymous@sos.domain. The UE can initiate the emergency session without a prior emergency registration for a home network user, or where the user is already registered. It would otherwise need to perform an emergency registration. The UE will try the emergency call in the CS domain, if possible. It will provide as much identity information as possible, such as the International Mobile Equipment Identity (IMEI), Mobile Country Code (MCC), Mobile Network Code (MNC), temporary tel URI, and location information.

The P-CSCF detects an emergency services request from the UE and differentiates it from normal registration processing. It processes the anonymous request and prioritizes the session. It determines the E-CSCF that will be responsible for handling the call, and routes the message to that E-CSCF. Upon receiving an emergency services request, the E-CSCF needs to determine which PSAP the call can be routed to. It may obtain the location information from the UE, if it has been populated. In most cases, however, it will need to request the location from the LRF. The E-CSCF may also forward the request to an Emergency Call Server supported by the National Emergency Number Association (NENA) VoIP architecture. The E-CSCF can route the call according to the established IMS and the PSAP configuration. It can route the call to a circuit-switched PSAP on the Mi/Mg interface via the Border Gateway Control Function (BGCF)/ Media Gateway Controller (MGC). It can route the call to an IP PSAP via the Mm/Mb interface.

The LRF assists in obtaining the present location information of the UE. The LRF relies on two important functions. The Location Server, such as the GMLC, SLC, or the MPC to request the position coordinates of the UE from a PDE network or the handset. The Routing Determination Function (RDF) to provide the information to the E-CSCF for routing the call to the correct PSAP. The RDF also provides the function to allocate an emergency service query key (ESQK) and a routing number for the PSAP to query the LRF. The initial request to the LRF by the E-CSCF is to obtain a low-accuracy location to determine the PSAP covering that geographical area. Subsequently, the PSAP can request a high-accuracy location to track the UE precisely. The LRF, RDF, and Location Server may be offered in a single element or in different physical configurations.

7.6 Chapter Summary

This chapter unfolded the exciting next-generation service paradigms that are possible to deliver with IMS. We examined the characteristics of multimedia services and noted the new models of bundled, converged, combinational, and converged services that are possible with IMS. We looked at presence, messaging, PoC, conferencing, and video sharing as examples of consumer services. While these also find applicability in the enterprise, we focused on the services enabling collaboration such as group lists and application whiteboard sharing.

To get a feel for blended services, we looked at IMS-IPTV services, location-based services, and video surveillance. We concluded by noting how emergency services, which are mandated by government regulation, can be supported by the IMS architecture.

CHAPTER 8

The Promise of Convergence

The emergence of new standards and innovative technologies offer a luring proposition to provide a new service that can be quickly monetized. The problem is that they quickly tend to grow into their islands and subsequently become difficult to maintain or they wither away. Convergence helps to overcome this problem and benefits by allowing progressive standards and technologies to leverage existing revenue-generating assets. Convergence can be realized at the network, service, or device level. Our focus is primarily on the network and service level to see how the IP Multimedia Subsystem (IMS) plays its role in convergence.

8.1 Converging IMS with Other Networks

We learned in Chapter 1 about the various standards that have been helping to shape IMS and acknowledge its presence as a core network in their architecture. We will now examine how the wireline, cable, and the next generation of wireless standards are evolving with the IMS core.

8.1.1 Convergence with Wireline Networks – TISPAN NGN

The European Telecommunications Standards Institute's Telecoms & Internet converged Services & Protocols for Advanced Networks (ETSI TISPAN) has been instrumental in defining the evolution of the fixed circuit-switched wireline networks as they move towards an all-Internet Protocol (IP) network. The architecture and principles for the all-IP Next Generation Network (NGN) standard are defined by ETSI TISPAN. ATIS has also contributed the North American set of requirements for the NGN. The initial focus of NGN was primarily for fixed networks. However, it has continued to evolve to the harmonized view of both fixed and mobile networks, as reflected in Figure 8.1.

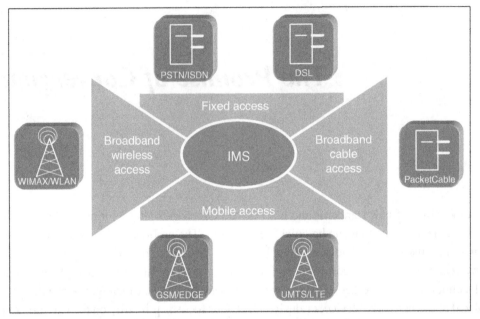

Figure 8.1 The NGN Converged Network

The NGN aims to provide a secure multi-protocol, multi-access, and multi-service IP-based network. It delivers multimedia services to both fixed and mobile terminals, enabled through a common core network. It supports the functions of nomadicity and mobility, between fixed and mobile, users and devices.

The TISPAN NGN embodies similar principles as the Third Generation Partnership Project (3GPP) IMS, which include end-to-end Session Initiation Protocol (SIP) signaling, core session control, and guaranteed Quality of Service (QoS). It is not access-agnostic as the 3GPP view, however, since its focus is on fixed/wireline access. The roadmap of the TISPAN NGN has defined two releases so far. The focus of Release 1.0 has been to provide a fixed broadband access to the core network aimed primarily at PSTN replacement and multimedia support. Thus the foundation has stressed more on the transport and QoS resource and admission control for these networks. This release integrates the 3GPP Release 7 IMS as the core network for session control and signaling for IP multimedia services. Release 2.0 focuses on aligning the IMS core fully to 3GPP Release 7 and higher. It also focuses on the service layer to provide Voice Call Continuity (VCC), Fixed Mobile Convergence (FMC), and TV Service over IP (IPTV) services.

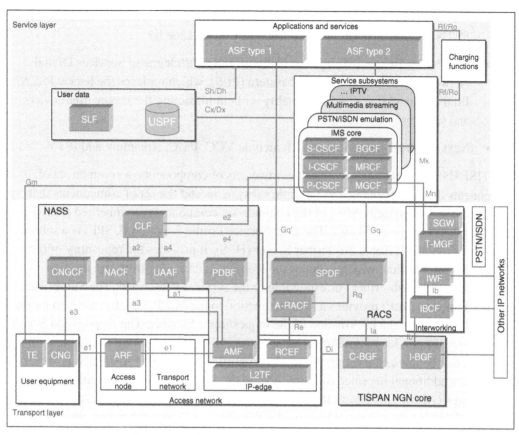

Figure 8.2 The TISPAN Release 1.0 Architecture

NGN defines a two-layered architecture that comprises the transport and service layers as, seen in Figure 8.2. Each layer consists of subsystems, which provide a modular approach to deliver functionality within that layer. NGN describes the transport layer with the functional entities that provide the IP transport access, connectivity, and control. The two principal subsystems in the transport layer that are responsible for providing the IP connectivity, are the Network Attachment SubSystem (NASS) and the Resource and Admission Control Subsystem (RACS).

The service layer provides the support for service logic, control, and delivery. The subsystems in this layer are:

• The 3GPP IMS core subsystem to deliver multimedia services.

- The PSTN/ISDN Simulation Subsystem (PSS), which provides a translation of PSTN features in an IP network, such as Calling Line ID.

- The Public Switched Telephone Network (PSTN)/Integrated Services Digital Network (ISDN) Emulation Subsystem (PES), which replaces the legacy PSTN/ISDN by emulating the functionality without impacting the legacy user devices and services.

- Next Generation Services, which include VCC, FMC, streaming and IPTV

The TISPAN NGN architecture introduces two sets of components: a common set of components that are shared across multiple subsystems and the set of components that are specific to a subsystem. Most of the common set components are inherited from the 3GPP IMS. These are as follows. The User Service Profile Function (USPF) is a subset function of the 3GPP Home Subscriber Server (HSS). It provides the repository of the user profile information, which includes the identity, address, security, and location information of the subscriber corresponding to the services subscribed. The Subscriber Locator Function (SLF) provides a similar function to the 3GPP SLF. Its role is to locate the USPF that stores the information of the requesting subscriber. The Application Server Function (ASF) provides a similar function to the 3GPP AS by hosting the application logic that can be invoked through the session control function. TISPAN NGN, however, provides an additional interface from the ASF to the Resource Admission and Control Subsystem (RACS). This allows the application server to control IP-flow resources. The Inter-Working Function (IWF) provides the necessary support between different SIP profiles and between SIP and H.323 systems. The Interconnection Border Control Function (IBCF) provides the edge functions specific to the service provider's network. It supports session border control, interfacing for bandwidth control with the RACS, and invokes the IWF as necessary. Finally, the charging and event collection functions support both online and offline charging similar to 3GPP IMS.

Most components within the core IMS subsystem provide the same nature of functionality as the 3GPP specifications. The set of the three Call Session Control Functions (CSCF), Media Resource Function (MRF), Media Gateway Control function (MGCF), and the Border Gateway Control Function (BGCF) provide the necessary functionality for session control and media handling. It also supports the same principles of the application service layer to provide Session Initiation Protocol (SIP), Legacy, and Open Service Access (OSA) services.

At a high level some of the differences with respect to the 3GPP IMS are as follows. The P-CSCF also supports the function of the Application Layer Gateway (ALG), which provides the support for traversing Network Address Port Translation (NAPT). It also interfaces with the NASS to obtain the physical location of the requesting subscriber. Some of the constraints imposed on SIP by 3GPP to support wireless infrastructure are not essential. SIP compression is not required. The SIP timers can be adjusted according to wireline network needs and the SIP preconditions are not mandatory. The Access-Network-Info Headers in SIP are extended to support fixed networks such as xDSL.

The transport functions in the TISPAN NGN architecture comprise elements that are specific to the application in a fixed broadband network. These comprise of the RACS, the NASS, and the transfer functions. The transfer functions comprise the termination and inter-working at Layer 2, Layer 3, and higher. The termination function for Layer 2 protocols includes Point-to-point Protocol (PPP) or xDSL. The Border Gateway Function (BGF) supports border control functionality at the IP packet level in the access, core, or inter-core network. The BGF supports the functions for IPv4/IPv6 conversion, NAPT and Network Address Translation (NAT) traversal, traffic screening, and topology hiding. The Layer 3 and higher protocol transfer functions are enabled by the MGF and the SGF for media and signaling, respectively.

The policy functions for resource admission and control are provided by the RACS. The architecture bears similarity to the 3GPP Policy architecture. The elements in the RACS also use diameter to communicate the policy-related information. The RACS provides the support for guaranteed QoS by resource reservation. It also supports relative QoS with diffserv. The RACS subsystem comprises the following elements. The Service Based Policy Decision Function (SPDF) communicates with the Application Function (AF), a P-CSCF for instance. The AF provides the resource requirements from the session descriptor. The SPDF can take the decision to authorize the QoS resources, which are communicated to the Access Resource and Admission Control Function (A-RACF) or the C-BGF. The A-RACF is located in the access network and is responsible for directing the SPDF decisions in the Resource Control Enforcement Function (RCEF). Depending upon the mode—Guaranteed QoS or diffserv—it sets the appropriate Layer 2/Layer 3 policies or diffserv markers, respectively in the RCEF. The RCEF is located in the access network and can apply the gating function on the Layer 2 termination. The Core Border Gateway Function C-BGF is an edge router located at the border of the core network. The SPDF can take the decisions to apply a gating function to the IP-flows at the C-BGF.

The NASS provides the set of functions for the initial IP-Connectivity of a fixed/mobile user endpoint (UE) to enable services provided by the NGN. This module is responsible for managing the IP address space within the access network and providing authentication to service sessions. Network attachment is provided based on either implicit or explicit user identification credentials stored in its database (respectively, physical or logical Layer 2 addresses, or user name and password).

The NASS comprises the following functional entities. The Network Access Configuration Function (NACF) is responsible for the IP address allocation to the user equipment and it may provide some additional parameters. This service can be provided by a Dynamic Host Configuration Protocol (DHCP) server. The Access Management Function (AMF) is an inter-working (translation, forwarding of user requests) between the access network and the NACF. The Connectivity Session Location and Repository Function (CLF) is used to associate the user IP address to the physical location information. The Profile Data Base Function (PDBF) stores the user profiles and authentication data. The User Access Authorization Function (UAAF) performs authentication for network access, based on the user profile stored in the PDBF. The CNG Configuration Function (CNGCF) is used to configure the Customer Network Gateway (CNG) when necessary.

8.1.2 Convergence with Cable Networks – PacketCable 2.0

For several years the hybrid fiber coaxial (HFC) has been the main transport of delivering video services to residential users. Cable service providers in recent years have successfully evolved this to deliver broadband data access for Internet connectivity. Going a step further, they have been able to extend the IP broadband access to Voice over Internet Protocol (VoIP)-based telephony. These standards for the cable industry have been provided by CableLabs. It pioneered the Data Over Cable Service Interface Specifications (DOCSIS), which defined the architecture for the cable television and data delivery.

PacketCable is the standards specifications from CableLabs for converged multimedia services for the broadband cable industry. PacketCable has been the evolution of the cable-related standards towards a SIP session-based architecture. PacketCable standards commenced from 1.0, which provides telephony support. This was delivered by using Network-based call signaling (NCS) and residential embedded Multi-Terminal Adapters (E-MTA). PacketCable 1.5 defined a SIP-based session support moving away from the NCS call control. The next release called PacketCable Multimedia, introduced service agnostic QoS and accounting functions. Finally, PacketCable 2.0 aligns the architecture with IMS Release 7 to support the SIP endpoints and SIP-based servers.

The broadband cable industry has been successfully delivering video, data, and VoIP telephony to residential customers. However, these services are provided as a bundle. PacketCable realizes the IMS promise of delivering enhanced services with both multimedia and communications components. It effectively enables a platform to deliver Quad-Play services, which can deliver a rich capability set than just bundling wireless/ cellular services. The architecture also enables using IP video, which enables the service providers to deliver video with both traditional cable signaling and IPTV.

PacketCable 2.0 enhances the current experience of broadcast or on-demand video to support integrated video communications. This includes video telephony, presence, and group list features. It also enables the integration of VoIP calling with the TV with features such as Call Indication and control from the TV. PacketCable 2.0 also defines mobility services and integration with cellular and wireless networks. This allows a Cable VoIP user to maintain a single directory number and roam seamlessly between the residential VoIP over WiFi network and a cellular network. PacketCable 2.0 also enhances streamed audio and video with guaranteed QoS. It also continues to maintain the features for SIP-based VoIP over cable networks, which can support traditional telephony and features such as emergency calling and lawful intercept.

As we see in Figure 8.3, PacketCable 2.0 defines a clear architecture to integrate with the IMS core network and deliver IP multimedia services. In doing so, the specification describes the functions of the IMS core network as we have seen, but the following areas define the principles to extend IMS to the cable network.

- The Cable Network Infrastructure as an IP-CAN for the IMS core network.

- The Edge functions to enable interconnect of the cable access network.

- Policy functions for bandwidth and resource control.

- Enabling the inter-working with the cable VoIP systems.

The PacketCable 2.0 architecture logically groups the various functional elements that interact with the IMS core network. These groups provide the Local Network (in the home), Access Network, Edge, Application, Interconnect, and Operational Support

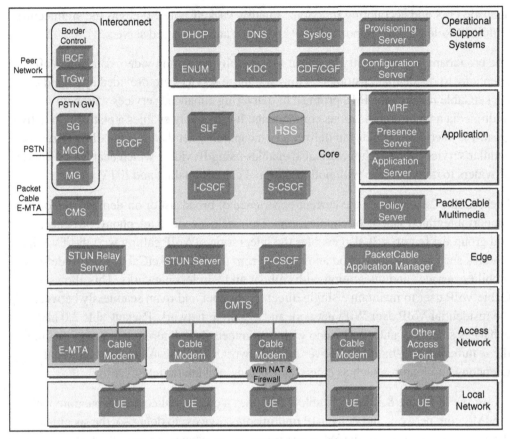

Figure 8.3 The PacketCable 2.0 Architecture

System (OSS) functions. We examine the broad features that distinguish themselves from the IMS architecture, which we have seen so far.

A PacketCable 2.0 user equipment (UE) has a public user identity. Like IMS, the public user identity can be associated with many devices. Since the cable user devices are normally fixed, they do not require the support of an ISIM. Consequently, they may not have a private user identity.

In the Access Function, the IP-CAN for a PacketCable system differs from the wireless networks. PacketCable 2.0 supports access through either DOCSIS network, non-SIP E-MTA, or a non-cable access network. The intent for a non-cable access is to allow a dual mode wireless device to obtain service from the network. It could use WiFi for the local network access by going through the broadband cable access, or use the wireless cellular network for network access. The DOCSIS access network comprises residential cable modems terminating to a Cable Modem Terminating Switch (CMTS). The CMTS performs an IP routing function between the Ethernet interfaces to the IP network and the Hybrid fiber/coaxial (HFC) connections to the cable modems.

The policy function for multimedia functions in the PacketCable 2.0 architecture is applied across the PacketCable Access Manager (PAM) and the PacketCable policy server. The PAM is responsible for the determination and management of QoS resources allocated for a particular session. It also determines the IP flows required for the session, and the bandwidth being requested for these flows based on the session descriptor that it receives from the P-CSCF. The PAM and the Policy server combined perform the role equivalent to the Policy Charging and Resource Function (PCRF) in the IMS reference architecture. The Policy server performs a proxy function. It communicates the QoS parameters to the CMTS, where the policy decisions are enforced on the IP-flows.

In the Edge function, PacketCable 2.0 also requires the use for Simple Tunneling of UDP through NAT (STUN) for supporting the media to traverse through NAT and firewalls. PacketCable 2.0 leverages the existing OSS framework for cable equipment and provides a suitable management framework for PacketCable UEs. The current framework for embedded UEs (E-UE) defines the connectivity to a provisioning server, by which the endpoints can be configured to obtain connectivity to a local network, discover the P-CSCF, and perform initialization functions. The framework also identifies Simple Network Management Protocol (SNMP)-based managers for alarm and management reporting.

8.1.3 Convergence with 4G Networks – WiMAX

WiMAX technology is based on IEEE 802.16 standard, and has been designed to provide wireless data over a long range comparable to cellular networks. WiMAX has evolved from point-to-point backhaul services to portability and nomadicity. It is also viewed as Fixed WiMAX and Mobile WiMAX, based on its ability to support handoff between base stations. It provides a significant advantage as a wireless last mile access for broadband access, as an alternative to Digital Subscriber Line (DSL). Supporting the Orthogonal Frequency Division Multiple Access (OFDMA) radio technology, it can deliver uplink and downlink data rates of up to 70 Mbps. It has thus been viewed as a contender for fourth generation (4G) wireless networks. The WiMAX standards for the conformance, interoperability, and inter-working are provided by the WiMAX forum. The networking group of the WiMAX forum has defined the inter-working with the 3GPP/3GPP2 standards, and has also proposed the mechanism for inter-working WiMAX with IMS.

The WiMAX network provides an IP-CAN to the IMS core network for the transport of IMS signaling and bearer traffic to support of IMS-based services to WiMAX capable mobile station/UE. The current WiMAX Release 1.0 recommendations provide support for IMS Release 6 and higher. At the time of writing the Release 1.0 standards focus WiMAX as a transport. Homogenization of policy, charging, and subscription features are not fully defined as yet. The use of WiMAX does not require any changes in the IMS core network or its interfaces. WiMAX mobile stations can host an IMS client to communicate with the IMS core network.

The WiMAX architecture comprises of three functional entities:

- The mobile device or the user equipment.

- The Access Service Network (ASN), which provides the radio access to a WiMAX subscriber. The ASN comprises radio base stations and one or more ASN gateways. The ASN is responsible for the functions of WiMAX radio link connectivity, enabling AAA for the subscriber sessions, network discovery, and mobility functions.

- The Connectivity Service Network (CSN) provides IP connectivity services to the WiMAX subscribers. The CSN is responsible for the IP address allocation for the device, Authentication, Authorization, and Accounting (AAA) services, policy and admission control, roaming, and other services.

The interfaces between MS, ASN, and CSN are identified as R1 and R3, respectively.

The WiMAX ASN can integrate with a CSN that supports IMS. As we see in Figure 8.4, there are three cases where the mobile is in the home network or the visited network, where the visited network may or may not support IMS.

One of the challenging areas that is yet to be defined fully, is to unify the QoS between WiMAX and the IMS network for resource admission and control. While release 1.0 defines only a method of pre-provisioned service flows, the architecture provides the mechanisms for controlling service flows and applying policy rules. The ASN function

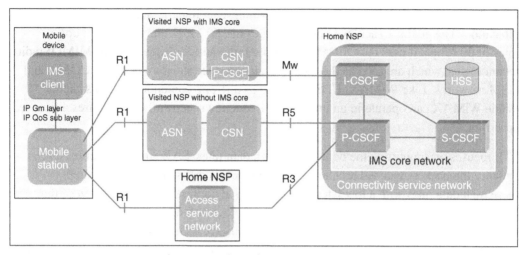

Figure 8.4 The WiMAX Access to IMS

supports local policy rules for service flow authorization. It also defines an interface to the policy function in the CSN to obtain the policy directives. The IMS PCRF acting as the policy decision function in the CSN would communicate the resource and admission control information to control the service on the IP-flows in the ASN. Mapping the QoS information obtained from SDP and 3GPP QoS classes to WiMAX classes is, however, required. WiMAX defines the following five QoS scheduling services - Unsolicited grant service (UGS), Real-time polling service (rtPS), Extended real-time polling service (ertPS), Non-real-time polling service (nrtPS) and Best effort (BE).

8.1.4 Convergence with 4G Networks – LTE / SAE

As we noted in Chapter 1, the standards from 3GPP have helped to evolve the wireless cellular network from a circuit-switched to a packet-switched core. IMS has been a part of this transformation as it provides the core network to deliver IP services. In order to deliver wireless IP-based services at high data rates requires efficiency improvements in the Radio Access Network (RAN) and the radio technology. With Release 7 of the 3GPP standards being the current baseline architecture definition, the future releases continue to optimize the aforementioned goal. This is accomplished by the Long Term Evolution (LTE) of the Radio standards and the System Architecture Evolution (SAE) of the network to support it.

LTE aims to provide higher peak data rates, which scale linearly according to spectrum allocation. The goal is to achieve downlink rates of 100 Mbps and higher and uplink rates of 50 Mbps. It uses the OFDMA/Multiple Input and Multiple Output (MIMO) radio technologies, which are seen as the common air-interface standard between CDMA and GSM networks. Like WiMAX, LTE is viewed as a strong contender for 4G systems. While WiMAX can operate in an unlicensed frequency spectrum, LTE requires a licensed spectrum.

LTE focuses on supporting the demand for high data rates. SAE aims to optimize and enhance the packet-switched architecture to deliver to increased IP traffic to support these high data rates with a low latency. SAE looks at addressing simplification of the IP network to support LTE and other access technologies. It addresses providing control elements to support mobility between heterogeneous access networks. SAE also strives to flatten the multi-tier RAN architecture and reduce the number of hops for reducing latency.

As we observed, LTE and SAE go hand in hand. LTE being a radio technology, does not directly impact IMS. The SAE shall, however, influence the edge functions and access to IMS. The standardization of LTE and SAE is still in progress at the time of writing this book. We examine the proposed view of SAE and its interconnect with IMS in Figure 8.5.

The SAE architecture is also referred to as the Evolved Packet Core System (EPS). The is capable of supporting 3GPP Radio Access, 3GPP IP access, and non-3GPP IP access. The 3GPP Radio Access comprises the current generation of RANs. These are the 2.5G GPRS and EDGE RAN (GERAN), the 3G Universal Terrestrial RAN (UTRAN), and the 4G LTE based Evolved RAN (E-RAN). The IP access network access can be provided by the 3GPP WLAN or other non-3GPP IP access. The SAE provides the functional elements to support both service access and service continuity between these multiple access networks.

Effectively, SAE addresses the architecture aspects of mobility with the evolved RAN, and inter-system mobility with the other 3GPP and IP access networks. It focuses on policy control and charging, and addresses the issues for roaming. The SAE defines two new functional groups: the Mobility Management Entity (MME) and the Inter Access System Anchor (IASA).

The MME supports the LTE-evolved RAN. It manages and stores the UE context, which includes identity and state security parameters information. It provides authorization

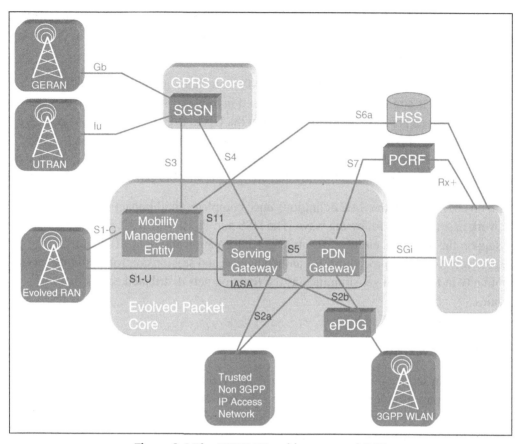

Figure 8.5 The LTE/SAE architecture and IMS.

for the UE to access the E-RAN or camp on a different access network. It enables user authentication and also allocates the temporary identities to be allocated to the UE. The MME is also extended to maintain the idle state UE terminations and is responsible for paging when downlink data arrives.

The IASA comprises the 3GPP anchor and the SAE anchor functions. Anchoring is the process of creating, joining, and maintaining two connections for the purpose of service continuity or mobility. The 3GPP anchoring is implemented by the Serving Gateway, which provides mobility between the 2G/3G access system and the LTE access system by anchoring the user-plane. The SAE anchoring implemented by the Packet Data Network

Gateway (P-GW) that provides mobility between 3GPP access systems and non-3GPP access systems, by anchoring the user-plane. The P-GW provides the access to the Packet Data Network and the IMS Core network.

The P-GW interacts with the P-CSCF in the IMS Core Network with over the SGi interface which is similar to the Gi interface used in the GPRS networks. This logical signaling interface has been kept similar so that the UE to P-CSCF interaction could use the GPRS or the EPS as the bearer. To support the interworking with IMS, the P-GW is required to support P-CSCF discovery, dedicated signaling bearers with QoS control and also provide the Gx interface for policy and charging control of the bearer for IMS flows.

With this background, the SAE definitions once complete, shall impact the IMS core network in some of the following ways. The HSS and PCRF functions shall need to support the new SAE entities for subscriber information and policy functions, respectively. Complex procedures for authentication and session control involving service continuity in IMS, shall get simplified as the function gets distributed across the SAE entities.

8.2 Service Convergence

As the telecom service providers grow their network to offer multiple revenue generating services, they are surmounted by a problem. The network ends up supporting a diverse set of applications and services hosted on heterogeneous platforms, which are often not seamlessly integrated. This results in the network consisting of multiple service silos, also referred to as a stovepipe approach, as depicted in Figure 8.6. These services include voice mail, unified messaging, fraud management to name a few. The problem continues to increase as service providers strive to bring in Internet and Web-related services, which introduces new silos for video, content, social networking, and so forth. To solve this problem, service providers have taken the cue from enterprise systems, which have faced a similar problem of multiple disparate applications. The solution has been to migrate towards a Service-Oriented-Architecture (SOA).

Since IMS is often seen as a unifying platform for converging applications, the practical realization of an IMS architecture in a telecom service provider network must have a clear strategy for service integration. There is a demand for IMS to conform in an SOA model, so that the converged services it brings do not result in standalone islands.

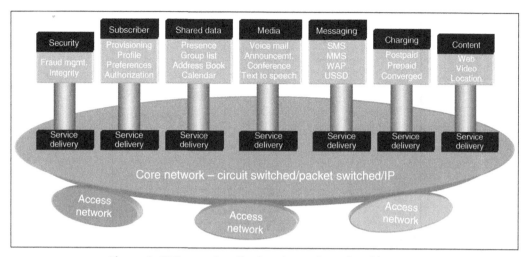

Figure 8.6 The service silo function-oriented architecture.

SOA is an approach to building an architecture that comprises distinct functions packaged as a set of services. These services can communicate with each other by exchange of data, and can be re-used across various business processes. The reason SOA is seen as a panacea is because it enables a cooperating model of services, which can be utilized in a repeatable manner and not result in a service silo model.

The nature of services in a converged network is also transforming from typical telecom-centric such toll-free services to becoming Web-based services. Web services have also provided a model that has gained acceptance for an SOA-enabling model. Web services utilize a service interaction architecture based on protocols that help describe a service interface, define the formats for the data exchange, and provide the mechanism for the transfer of data.

Web services are based on three types of entities, which we note in Figure 8.7. A *Service Provider*, which delivers a set of functions packaged as a service. The Service Provider has a well-defined interface described in the XML-based Web Services Definition Language (WSDL). It publishes its interface to a Service Registry. The *Service Registry* exposes this data in the Universal Description, Discovery and Integration (UDDI) protocol. The *Service Requestor* discovers the service interface information from the Service Registry, and can invoke the services using a Simple Object Access Protocol (SOAP)/Hypertext Transfer Protocol (HTTP) transport protocol. The Simple Object

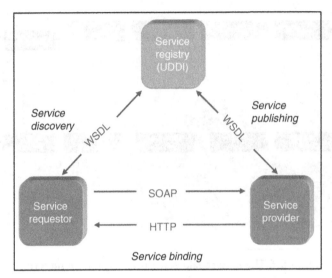

Figure 8.7 The basic web services model.

Access Protocol (SOAP) provides an XML-based message format with bindings to the underlying HTTP protocol.

The Web Services Interoperability organization (WS-I) is an open industry organization that provides the definition of the Web services interoperability. The WS-I defines a basic profile for Web services, which comprises SOAP and WSDL. It defines the basic security profile for transport security and SOAP messaging security for the considerations of the basic profile.

While the method described to access Web services is simple and elegant, it is does not adequately cover the complexity of service invocation and interaction with non-Web services. The resulting architecture with this method requires a number of adapters and resembles a hub and spoke. This is why Web services by themselves are not seen as an SOA.

The concept of a service bus offers a better architecture construct that provides the communication medium for the SOA. The service bus is the set of interface invocation, mediation, and routing functions essentially built over messaging middleware that hides the complexity of a service interface to a service requestor. The logic supported by a service bus supports the implementation of business rules to create new services by brokering and aggregating multiple services. In addition to the advantage of reducing a point-to-point service connection, the service bus abstracts the interfaces of heterogeneous service elements. This helps in reuse of these services without the need of implementing new adapters.

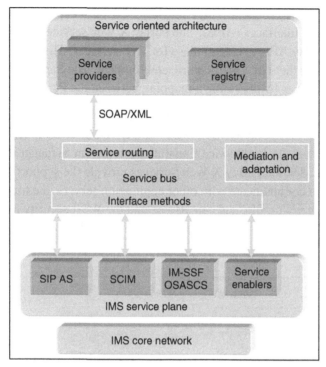

Figure 8.8 The service bus model for SOA.

The directions for integrating IMS services into the SOA are being provided by three different approaches, which we will examine shortly. The general concept is illustrated in Figure 8.8. The essence is to integrate the elements in the IMS service plane to align with a service bus model. An abstract view is shown for the service plane elements to use the interface methods provided by the service bus. The service bus is designed to provide the service routing/mediation and adaptation and interfaces with the SOA using SOAP/XML/WSDL interfaces. The logical concept of the service bus can be realized by the following approaches.

The OSA/Parlay Approach

The OSA/Parlay approach is based on the principle of network independence. The OSA utilizes a Parlay gateway, and the gateway comprises two parts: the Framework and the Service. The Framework provides the ability for the network access, routing, and terminating functions to service user or the service provider. The Service part comprises the set of interfaces to the applications within the network, such as call control, mobility, messaging, presence, and so forth. These services have been exposed with the Parlay Application Program Interfaces (APIs). The Parlay X Web services provide a further abstraction over Parlay APIs to extend these for Web services. These also simplify the Parlay APIs to utilize them in a Web-centric environment. The Parlay Gateway can deliver the Web Services by providing a mapping between the Parlay X and Parlay APIs.

The Parlay X Web Services provides a possible solution of the service bus concept. The Parlay X Web Services support WSDL, SOAP and UDDI discovery. The Parlay Gateway also provides a Service Broker/Service Capability Interaction Manager (SCIM) function. As shown in Figure 8.9, the Parlay X Web Services provide the support for the IMS enabler functions to communicate with other application servers in a trusted domain. It supports the WS-I basic profile, and can also extend the WS-I security and WS-I policy profiles to the enablers, which require these functions in addition to the basic profile.

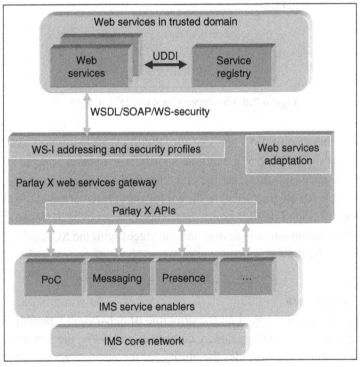

Figure 8.9 The Parlay X Web Services model.

The OMA approach

The Open Mobile Alliance (OMA) provides standardization for interoperable mobile services enablers. OMA has focused on defining an environment for creating and deploying services. It has defined a paradigm, OMA Service Environment (OSE), for an integrated service architecture. In addition to the definition of the OSE, OMA has also defined the integration of the OSE as a suitable service-enabling environment for IMS and OSA/Parlay.

The OSE model is seen as a progression in the telecom service architecture. As shown in Figure 8.10, traditional wireless services have been built with the IN and CAMEL Services Environment. These have been suitable to support invocation of services via transaction-oriented messaging. OSA/Parlay provided a higher abstraction of service

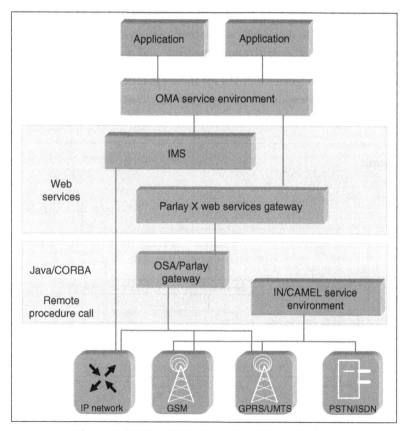

Figure 8.10 The OMA view of service evolution.

logic, which was independent of transport and transaction processing. OMA sees IMS and the Parlay X as elements to enable integration with multimedia and Web services.

We observe in Figure 8.11, the OSE comprises of a set of service enabler functions and their interactions. *Service enablers* are standardized components that package the specification, behavior, components, and interfaces required for a particular service function. For instance, Presence, Group Management, and PoC have their respective service enablers. An enabler can have an implementation within a terminal domain, i.e. UE or in the service provider network. The enabler thus has an implementation specific to the execution environment. Each enabler must provide one or more public interfaces. The specification of the interfaces is specified both as independent of the implementation and also providing language-specific bindings. The *service bindings* allow applications to access the enablers with a specific language interface such as Java or C++. The *Policy Enforcer* is an OSE element that applies policy mechanisms to the access to resources within the OSE. The policy rules are enforced for user preferences, authorization, charging, and privacy.

There are four interface categories defined for these OSE components. The I0 interface defines the public interface exposed by the enabler function for the application to invoke

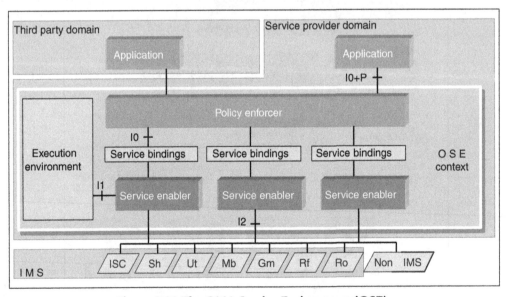

Figure 8.11 The OMA Service Environment (OSE).

its services. The I0 interface is used when policy rules do not need to be enforced for invoking these services. The I0+P interface identifies the category of interface functions, which need to have policy rules enforced on them. The I1 interface defines the interface between the enabler and the execution environment. The I2 interfaces describe the interfaces to the core network elements or the OSS/BSS functions in the network.

The OSE further details two specific functions and an additional interface for the IMS interaction, as shown in Figure 8.12. In the context of integrating with IMS, the OSE defines the Enabler Terminal Function (ETI) and the Enabler Server Implementation (ESI). These are executed in the UE and the application server environment, respectively. The OSE exposes the I0 interfaces between the ESI to applications and the ETI. The I2 defines the interfaces to the core IMS network, which are the standard IMS interfaces to the C-SCF, SLF/HSS, and Charging Servers.

The OSE defines Ut, a new interface between the ESI and the ETI, i.e. the UE Client and the application server. The Ut interface provides the UE client with the support for configuring

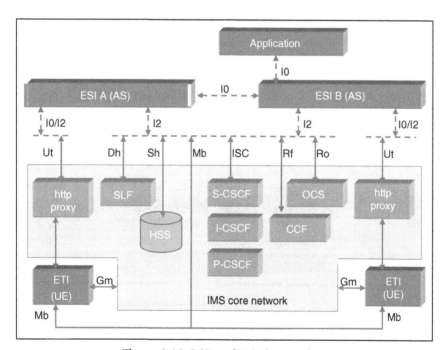

Figure 8.12 OSE and IMS interaction.

and updating user-specific data on the ESI. The user-specific data relates to presence lists, group lists, and authorization information. The Ut is an XCAP-based protocol.

The SDP Approach

The support infrastructure of the telecommunications network, also referred to as the OSS/BSS, supports critical back-office functions for the successful delivery of telecommunication services. Traditionally this has been implemented as a process-oriented architecture. It modeled the business functions and processes into their respective process architectures. The approach of a Service Delivery Platform (SDP) evolved to support these back-office functions in a service-oriented rather than process-oriented architecture.

The SDP was initially conceived to ease the content delivery for ring tones, games, and other downloads, and reduce its dependence on intensive back-office functions. The concepts behind the SDP are as shown in Figure 8.13. In the absence of standards for the specifications of the SDP, the OSA/Parlay and the OMA are considered a good foundation for the network interface technology. The SDP provides a service execution environment for executing the business logic of the back-office services it supports. The interaction and invocation of these services is handled by an SOA service orchestration

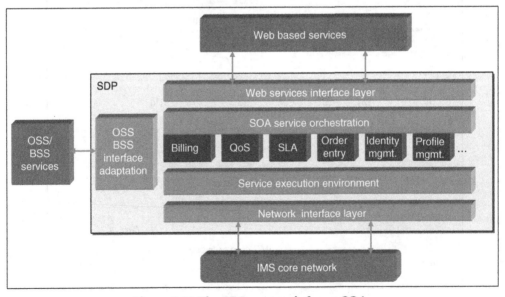

Figure 8.13 The SDP approach for an SOA.

function. The role of this function is to organize the service access and provide the necessary sequencing and mediation between the services that would provide the business logic for the execution of a service request. The SDP supports the necessary interface adaptation to both Web-based and other OSS/BSS services.

8.3 IMS and Web 2.0

In the references so far, we have used a broad term of Web services. We now distinguish these services to reflect the current trends. The first generation of Web services provided a platform to publish data through Web servers. Users accessed this data using HTML/HTTP and XML-based technologies in browsers. The second generation termed as Web 2.0, focuses on enhancing the usability of this initial Web platform. This is achieved by several methods. At a communication level, Web 2.0 promotes services with social networking, collaboration, recommendation, blogs, participation, and wiki. At a medium level, it supports user content video, podcasting, and broadcasting. At the technology level it harnesses Asynchronous JavaScript and XML (AJAX), eXtensible Hypertext markup language (XHTML), widgets, Ruby, and SOAP. Mashups play an important role in Web 2.0 to unify these methods. Mashups that are a Web application, can combine data from more than one source and present it in a unified view. This offers a paradigm in Web 2.0 to create new services by a blended application.

Although standardization and open APIs has been a goal for Web 2.0, there are still no clearly defined standards for building and using Web 2.0 applications services. How does IMS fit in the Web 2.0 model? In principle, Web 2.0 services still function with the underlying Web server model and are also moving towards SOA capability.

One point we must recognize is that delivering a Web 2.0 application to a UE can be done as a data application, it does not need an IMS network. The true value that IMS delivers to a Web 2.0 application is in its ability to provide a converged service. This can be termed as an IMS-Web mashup or also referred to as a Telco mashup. An IMS-Web mashup is an extension of the Web mashup concept. This leverages data from a Web application and an IMS application and delivers an aggregate of these applications as a unified service. The unified service can be presented as a Web application or as a multimedia application delivered to an IMS endpoint.

Let's consider an example to make this distinction. An IMS service such as messaging, presence, group management, or push to talk can be integrated as an application in a social networking application. This allows a Web 2.0 application user to invoke an IMS service and communicate with an IMS user. Alternatively, an IMS user can click for an option to display the weather forecast while viewing a football game. The data can be presented as a mashup with the video stream.

To explore this further, we examine a popular technology for Web 2.0. The use of Asynchronous Javascript and XML (Ajax) has provided several Web 2.0 applications with better usability. Ajax is an elegant approach to giving a dynamic look and feel to the Web pages. The entire Web page does not need to be reloaded whenever data is fetched from the Web server. This is visible in Google maps and the Yahoo portals. As the name suggests, Ajax uses the javascript function calls to request data from the Web server asynchronously. This allows extra data to be fetched and updated without the need for refreshing the entire page.

The effort to bridge an IMS application server with an Ajax API requires a bridge or an adapter function, as described in Figure 8.14. Java is a commonly available interface for

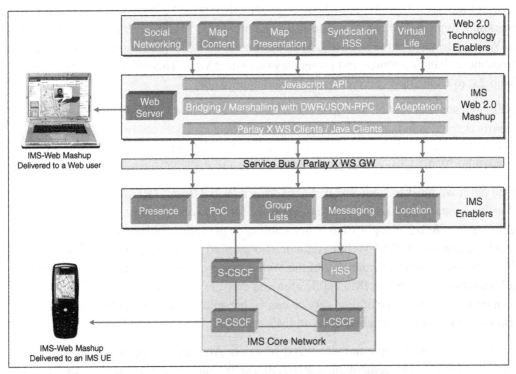

Figure 8.14 Converging IMS and Web 2.0 services.

most IMS, SIP, and AS, and also with the Parlay X WS implementation. The adapter, however, is required to provide the Java to javascript communication. Direct Web Remoting (DWR) and Javascript Object Notation Remote procedure call (JSON-RPC) are two methods that provide the marshalling of Java objects to javascript and allow the javascript to work with server side Java objects. This helps an API exposed as a method in a java class to be invoked remotely, providing a method to invoke services of an IMS application server/enabler from the Web browser through the mashup.

Once the IMS-Web mashup has been blended, it can be delivered through a Web server to a Web user or delivered to an IMS UE. A second tier of service blending at the IMS layer with the SCIM, can further result in enhanced services.

8.4 Chapter Summary

IMS plays a central role in the convergence of other networks. TISPAN NGN identifies IMS as the core network in the wireline broadband access network. The PacketCable 2.0 standard defines a similar view of the IMS as the core network for broadband cable access networks. Emerging 4G mobile broadband standards of WiMAX and LTE follow the similar paradigm of defining their role as an access network with IMS as the core network. This benefit of network convergence with IMS is possible by the access-network agnostic principles in its architecture. QoS, network admission and bandwidth control are essential to address in an IP network. These are also specific to the access network. While the TISPAN and CableLabs standards have effectively focused on the QoS aspects, WiMAX and LTE are still evolving to address these aspects and the inter-working of QoS with the IMS core.

Service convergence with IMS focuses on the paradigm to a service-oriented architecture. This model is essential to allow IMS services to fit in the transformation of the OSS/BSS and other Web services, from function-oriented to service-oriented. Integration with Web Services and SOA are being driven by the OMA and SDP models. Bridging IMS with Web 2.0 services is opening up a new dimension of converged services.

CHAPTER 9

Implementing IMS Functional Elements

The specifications of the IMS have focused on defining the architecture, the functional elements within it, and their interaction. The standards do not mandate or dictate a particular implementation to realize these functional elements. One route, which initial implementers have taken, is to successfully reuse existing network assets and evolve them to conform to the IMS reference architecture. For instance, SIP servers have been leveraged for session control and the HLR subscriber function was enhanced to support the HSS features. We will, however, focus on engineering these solutions from commonly followed paradigms and trends.

9.1 The Network Itself

To start with, let's understand the foundation of the service provider's IP network commonly in deployment. In recent years, the network backbone of the MNO became quite fragmented with disparate technologies of Asynchronous Transfer Mode (ATM), Frame relay, and Time Division Multiplexing (TDM) networks. Implementing an IMS network within a foundation of these domains adds to the complexity. Fortunately, a growing number of service providers have adopted IP/ MPLS as the technology for their network backbone to align with the 3G network evolution. Implementing an IMS network becomes more structured with the underlying IP/MPLS backbone. The IP/MPLS network offers ease of introducing new classes of IP services with differentiated QoS and traffic-engineered routing. This makes the implementation of the IMS network simpler.

MPLS is seen as complementary to IP. MPLS supports a connection-oriented framework upon a connectionless IP network. This provides the foundation for reliable QoS and bandwidth utilization similar to an ATM network at a lower cost and processing overhead.

MPLS uses labels to forward traffic within the IP/MPLS backbone. When IP packets enter the IP/MPLS-enabled network, MPLS labels are added between the Layer 2 and the Layer 3 header in the packets. This label is then used within the network to route the message to the next hop. When the labeled packet arrives at a label switch router (LSR), the incoming label is used to determine the path of this packet within the IP/MPLS network. The MPLS label is removed when the packet exits the IP/MPLS domain. The MPLS backbone can be configured to accept Layer 2 virtual LAN (VLAN) traffic by configuring the label edge routers (LERs) at both ends of the MPLS backbone.

The main advantage the labels provide is their ability to group or classify packets based on attributes or forwarding classes. Packets belonging to the same forwarding class get similar treatment, providing QoS for varied traffic types. The label contains information about the destination, precedence rules, and QoS information. This MPLS lookup and forwarding system allows for explicit control of routing, based on destination and source address, allowing easier introduction of new IP services. The routing of the packets is set up based on the traffic engineering supporting services such as a fast reroute in case of failures.

Starting with the layout of the network, which is the IP/MPLS backbone, the next steps for designing the IMS network are dimensioning or the capacity considerations, QoS planning, and security. The dimensioning of the capacity of each network element is based on a subscriber forecast, which translates into a traffic usage model. The traffic characteristics for the IMS network are estimated in terms of the average session durations, busy hour attempts, codecs, and multimedia data traffic based on the services supported in the network. These figures then provide the figures for the number of subscribers provisioned in the HSS, the throughput in terms of SIP sessions, and the message rate to be supported by the Call Session Control Functions (CSCF). This also results in identifying the number of instances of each functional element that is required, i.e. the number of S-CSCFs or the P-CSCFs that are required to support the traffic in the network.

In order to design the right level of service experience, the metrics used to define QoS at the packet level are delay, jitter, packet loss, and echo. The Mean Opinion Score (MOS) provides the Quality of Experience (QoE) metric. The codec selection is essential to deliver to these QoS levels. The MOS score ranges between 1 (for unusable) and 5 (for excellent). The MOS score for a Voice over IP (VoIP) call is between 3.5 and 4.2. The AMR codec offers a MOS score of 4.2 vs. a 3.98 score for a G.723.1 codec.

Given the security considerations, we have examined earlier, the security perimeter is being fortified by some of the following methods. The session border controllers are playing a significant role in insulating the core network from the public Internet. Resident functionality on these is used to prevent Denial of Service (DoS)/Distributed Denial of Service (DDoS) attacks, flooding, and spoofing. Firewalls with Internet Protocol Security (IPSec) support are placed at strategic points for secure network access. The traffic in the IMS network is established through a Layer 3 VPN in the IP/MPLS backbone, to be insulated from the other services provided by the service provider.

9.2 Implementing UE Applications/IMS Clients

Several handheld devices, third generation (3G) and smart phones, support an operating environment with open interfaces or Application Program Interfaces (APIs) to create new applications. This provides a suitable platform for creating client applications. This paradigm has been suitably extended for creating SIP-based clients to support some of the initial IMS applications such as PoC. However, this method has its limitations in supporting a larger set of blended applications across different terminal types. These features in a UE are integral to realize the benefits of the IMS network. The general characteristics of building an IMS application on a UE are as follows.

- The UE must be able to deliver a rich user experience. This implies an easy-to-use interface that can enable the applications that we discussed earlier

- The UE should be able to aggregate and assemble the support for various applications seamlessly to the end user.

- The UEs should be able to support the wide set of IMS protocols; SIP, Session Description Protocol (SDP), real-time protocol (RTP), Real Time Control Protocol (RTCP), XML Capabilities Application Protocol (XCAP), Message Session Relay Protocol (MSRP), Extensible Markup Language (XML), Hypertext Transfer Protocol (HTTP), and a set of service enablers.

- In order to support a variety of terminal equipment, the UE applications must be able to reside in a set of host operating environments or alternatively extend a portable interface.

- Due to the embedded nature of most terminal equipment devices, such as wireless phones, the UE applications must be compact and provide a small memory footprint.

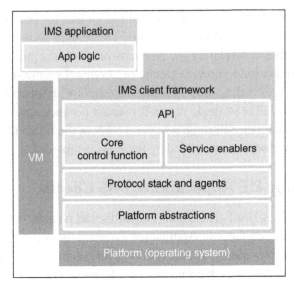

Figure 9.1 IMS client framework layered architecture.

These requirements call for a suitably layered and modular architecture. A value-added layered approach with an extensible IMS client framework in the UE provides the necessary foundation. This offers a common set of functions to build IMS client applications, as we can see in Figure 9.1.

This approach helps the IMS application developer to focus mainly on putting together the application logic. The IMS client framework abstracts the generic infrastructure functions of the protocol stacks, service enablers, control functions, and platform abstractions. An abstraction layer is essential in order for the IMS client to be portable across different mobile environments (e.g., JME, Brew, Linux or Windows-Mobile). The abstraction layer provides the mapping of the services provided by the host environment to the IMS client. The terminal platform will provide the virtual machine necessary for a Java-based client. The next layer is the set of protocol stacks that are required by IMS and their necessary client side agent functions. At the next level is the common set of functions required for communication with the IMS core network and the set of service enablers. Finally, the interface layer provides the programming interface or the Application Program Interface (API) to the user applications to invoke the service abstractions from the framework.

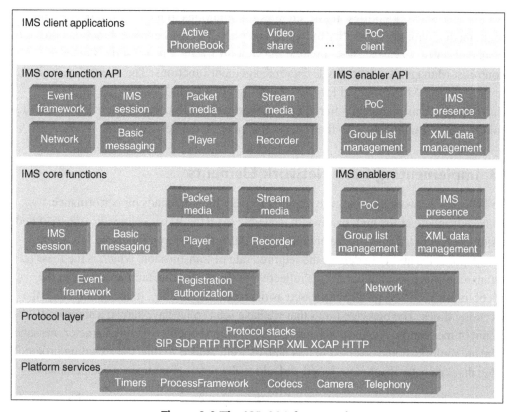

Figure 9.2 The JSR-281 framework.

These concepts have been expanded into the JSR-281 specification. While the specification focuses on providing the API in the form of java classes to the applications, its underlying model is flexible to be adapted to other environments such as Brew. The model is shown in Figure 9.2. JSR-281 extends the concept of the IMS client framework to provide an abstraction of a set of services, for building client applications. The client framework is itself an application that can be downloaded into a device supporting the Java virtual machine. It can then provide the necessary protocol and agent-level support for building an IMS client.

This model splits the control layer and the API into two sets. The IMS Core API provides the functionality for both the signaling and media flow to the IMS Core network. This enables the support for registration, authorization, event control, and session management with the core network entities. It also enables the functions to handle both packet and stream media, with the capability to record and play. The IMS enabler functions provide the second set of services. These include the support for the Push-to-talk, Presence, Group List Management, and XML data management functions. The client application can thus use the APIs provided for these building blocks and focus on the application behavior only. This results in a tiered model, where the IMS core, IMS enablers, and the applications communicate with their respective peer on the network side.

9.3 Implementing Core Network Elements

The IMS core network elements are required to meet high demands on performance. To take advantage of the high bandwidth available from the IP backbone, the transport and session-control functions must be engineered to deliver both throughput and low-latency. Throughput is the ability to deliver the required rate of transmission, which is the messages or transactions per second. Latency is reflective of the delays that are introduced in the path of transmission. Typically expressed as the round trip latency, it does not include the processing times. The message flows that we have seen so far have shown a significant amount of messaging. A typical session setup can take about 40 SIP and diameter messages. There are multiple hops traversed in this path as well. Added to this, the signaling protocols within the core network are text-based and add to the processing times at each element.

It therefore becomes imperative to engineer these elements with appropriate hardware and software. In addition to performance, the core network elements must also be able to provide scalability and high-availability.

9.3.1 Hardware Considerations

The attributes of high performance to support high throughput with minimized latency are a combination of the following: high performance packet processing, Input/Output (I/O) bandwidth, Central Processing Unit (CPU) performance, high-density, low latency memory, and efficient power usage. High performance packet processing is applicable to the edge IMS functions or the access network elements, which require special purpose hardware capable of supporting high-speed Gigabit Ethernet switching and I/O. High I/O bandwidth can be provided by a terabit backplane and a multi-gigabit Ethernet throughput per processing element. Techniques such as non-blocking help to reduce the latency. High-density enables scalability and efficient use of space for clustering.

There are two processor options for IMS core network elements. The first is a traditional set of high-performance processors employing Ghz frequencies and supporting instruction-level parallelism with preftch and caching techniques. The second is a set of multi-core processors, some of which also employ multithreading at a hardware level. Multi-core processors also address the disparity between memory and processor speeds. Memory access latency is seen as a disconnect, which undoes the performance gains of high-speed processors. The CSCFs in particular are expected to support the set up, hold, and tear down of a large number of concurrent sessions. The traditional processors are seen to fall short to exploit this level of concurrency. Multi-core processors and the support of chip-level multithreading allow the implementation of multi-threaded software to represent the concurrent sessions to truly exploit the parallelism.

The features of performance, scalability, and high-availability are inherited from traditional telecommunication systems. Commonly, these were implemented through proprietary hardware or by stacking form 1U (1.72 inch high) servers in a shelf. Today there are two industry standards that offer a better alternative with a blade server-based solution. A blade server or a blade for short represents a single board computer (SBC).

The Advanced Telecommunications Communications Architecture (ATCA) defines a shelf-based solution for the blade server-based architecture. This is based on the PICMG 3.x specification. Each shelf can accommodate a 14 blades chassis in a 19″ rack. Each 14-blade shelf is formed as 12U rack units. The ATCA backplane provides point-to-point connections between the blades and does not use a data bus. The ATCA provides a 10/100/1000 Base-T Ethernet Base interface to each board. It also provides a Gigabit Ethernet or a fiber channel switching fabric for higher bandwidth. The ATCA shelf also provides intelligent software as a shelf manager that monitors the performance of the blades. ATCA provides a CPU blade hot swap capability, which is the ability to replace the blade while the system is in operation.

The Blade Center is also a high-density chassis for hosting a set of blade servers. The chassis supports full hardware redundancy (power supply, I/O modules, management modules, L2 switching, mid-plane, and so forth), thereby minimizing single points of failure in the solution. It finds applications for both telecom and data centers. It also supports similar features and functionality, including fault-tolerant capabilities, hot-swappable redundant DC or AC power supplies and cooling, and built-in systems management resources in a 20″ deep chassis.

Figure 9.3 The ATCA shelf (Source HP).

Figure 9.4 The blade server platform (Source IBM).

Both hardware architectures have gained significant momentum. While they have their differentiation in terms of physical dimensions, cooling, density, and power consumption, they offer a similar physical model to implement the logical view of an IMS network element. The logical concept of an SBC/Blade is common to both systems. This makes it easy to model the system functionality and host the software elements on one or more blade servers and also provide scalability.

Most Blade server implementations provide high-performance symmetric processor architectures. In addition, the support of multi-core processors and standard memory configurations of 8–16 GB, provide a suitable platform for session control and transaction-oriented processing.

9.3.2 Software Considerations

Similar to hardware, the software-operating environment for building telecom network elements has evolved to industry standard choices from proprietary implementations. UNIX-based variants of Linux and Solaris have been able to deliver high-performance and reliability in the network. This has also been vital in the maintainability of these systems. Linux and Solaris provide a suitable environment for delivering the requirements of the IMS network elements. Carrier-Grade Linux (CGL) variants of Linux, which support kernel hardening embedded environments, offer additional support for reliability and performance. Most control plane software is developed in C/C++ for performance reasons. Having an interpreted environment adds to latency. This is why Java has not gained much acceptance in the control plane elements. On the other hand, the ability of Java to inter-work suitably with Web servers, has made it a suitable choice in the application plane. However, the advent of chip-multithreading has shown potential to improve the parallelism with the Java virtual machine, which may boost its performance beyond application servers.

The IMS network signaling elements are built on protocol implementations that we saw in Chapter. 4. The fundamental IP transport stacks Transmission Control Protocol (TCP), User Datagram Protocol (UDP), Stream Control Transmission Protocol (SCTP), and even the security protocols such as IPsec, are bundled with standard operating systems. Most core network elements also require a high-performance and conformant implementation of the SIP and Diameter protocol stacks. At a minimum, this is the implementation of the core SIP and Diameter RFCs. This includes the codecs message encode/decode capability, protocol state machines for handling dialogues, transactions, and sessions.

In addition to the base SIP RFC 3262 and the adjunct sets, the support for P-Headers is essential. Similarly for Diameter, the implementation requires the implementation of the base protocol RFCs—3588, 3589 with the support for working in the diameter server or client mode. In addition, the support for the various applications Cx, Dx, Rf, Ro, Rx, Gx, Tx, Ty, and so forth corresponding to the network element are also required.

9.3.3 Operational Considerations

IMS core network elements must be able to deliver to carrier-grade requirements. What this implies is that these elements must be able to perform reliably under stringent requirements of availability and serviceability. The typical requirements are for the system to deliver an availability of 99.999%. This measurement of 5-nines refers to being able to provide service with no more than 5 minutes of downtime in a year. This can be achieved by reliable hardware, hardened operating system environment, and exhaustive testing to a large extent. However, regardless of how robust the component is, there is always a chance of its failure. The key is to define a solution that eliminates any single point of failure. In other words, build a redundant architecture, where a backup system can take over for the failed system and avoid service disruption. Redundancy is a principle that has been used in traditional telecommunications networks infrastructure and is also required for the next generation networks.

In IMS network or in general, redundancy can be applied at the network level or at the element level. The difference is where the failure is detected and the failover is handled. Redundancy can be built into the network architecture to eliminate transaction loss in case of the failure of an element within the network. This means multiple instances of an entity performing the same function. For instance, if an application server instance is down, the S-CSCF shall be able to route the request to an alternative AS.

At the element level, redundancy can be applied for the full node or additionally to certain components, such as disks or CPUs. Replicating the complete element has its advantage of eliminating or minimizing failover and recovery time. A load-shared peer or a hot standby can provide service continuity in the event of a failure of the element. A cold-standby mode can minimize service disruption where the downtime is restricted to the failover time.

The redundancy paradigms are based on the goals for the failover time, cost, and scalability. In a 1 + 1 cluster configuration, a peer node offers a load-shared, hot, or cold standby model. The load-shared and hot-standby models require state and context replication of the sessions. In a load-shared model, the active/active policy assigns both nodes to be processing traffic and synchronizing their state and session context information. In a hot-standby model, the standby node is updated with the session context, but does not actively process any traffic, unless the failover is directed. The n + m cluster applies this policy to *n* active elements and *m* standby elements. Redundancy with load sharing in a n + m model helps to increase the capacity of the system, and is useful for stateless servers. It must be noted that state information replication and synchronization comes at a cost of performance.

The IMS core network elements also need to support the functionality for Operations Administration and Maintenance (OAM) functions. The methods are continuing to evolve in Release 8. These reflect the ability for the element to be managed as a unit or as part of a network or cluster. The essential management functions cover fault, configuration, accounting, and performance (FCAP). Fault management extends from the reporting of basic events and alarm conditions, to a central operations facility where these can be monitored, diagnosed, and corrected. Advanced methods include automated fault isolation and correction mechanisms. Configuration methods encompass the necessary provisioning required for the element to perform. Addressing, connectivity, and routing are basic configuration parameters common to most IMS network elements. Software version revisions and updates also are applied as a part of the configuration. It must be noted that most of the configuration updates are performed dynamically from a central management platform. Performance indicators collected as a set of measurements or key performance indicators, provide the data to assess the system as it performs under conditions of load in an actual field environment.

9.3.4 Implementing Session Control Elements

The IMS Session Control Functions are required to deliver a signaling intensive, high-performance, and highly available infrastructure for establishing and maintaining several thousand concurrent multimedia sessions. The CSCFs can be viewed as specialized SIP servers. Their functions, however, require more complexity in their ability to handle transaction sets in a User Agent (UA), Proxy, Registrar, or Back-to-back User Agent (B2BUA) role. In addition, the CSCFs also host their respective functionality for integrity, interrogation, leg management, and so forth. Finally, the CSCFs also play the role of a diameter entity for the exchange of subscriber, policy, and charging information.

The nature of implementation of these functions has been vendor-specific. As these elements are considered performance intensive, Java-based reference implementations have not gained acceptance. We explore these functional elements to understand better how they can be implemented.

Each of the CSCFs requires a base component of a SIP and a Diameter-enabling platform.

9.3.4.1 The P-CSCF

The role of the Proxy-CSCF as a network edge function and also the entry point in the roaming network, combines several functions:

- Proxy Function to support the registration process

- Security, Firewalling, and Network Address Translation (NAT)

- Policy

The P-CSCF function maintains state information as a stateful SIP-Proxy server. It therefore must be able to support a redundancy paradigm of $1 + 1$ so that a backup instance can continue to support the session setup in the event of a failure.

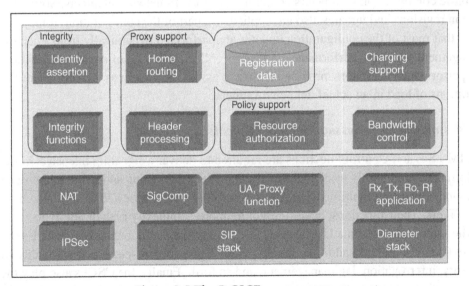

Figure 9.5 The P-CSCF components.

The Edge functions of the P-CSCF require it to provide a firewall capability and NAT support. Physically, these functions can be implemented in a separate blade server. In the initial implementations, these functions have been performed by a session border controller or an IPSec router.

The P-CSCF SIP signaling functions include the functionality that is performed after the P-CSCF has received an inbound session request. These functions:

- Assert the identity of the users and confirm this for subsequent processing for the S-CSCF

- Perform the Record Route Header to establish the P-CSCF in the path for outgoing messaging

- Maintain a record of the registered data that it has obtained in the exchange of signaling with the S-SCSF, in memory storage.

- Apply charging vectors

- Header processing for visited network ID and path headers

- Determine and perform the Routing to Home Network

The P-CSCF handles the SigComp functions that are invoked for compressing SIP messaging for over-the-air interface. The P-CSCF also hosts the policy functions for bandwidth control. This requires the logic for obtaining the session description and requesting the bandwidth and IP-flows from the PCRF. The P-CSCF conducts this with the diameter stack and the Tx/Rx message set.

9.3.4.2 The I-CSCF

The role of the I-CSCF is to locate and route to inbound signaling from a home or visited P-CSCF to the S-CSCF. It is also responsible for applying network domain security for topology hiding. To support these functions, the I-CSCF functions as a redirect server and its role is primarily as a stateless server. It can deliver reliability with a redundancy paradigm of N + 1, so that any backup instance in a cluster can continue to support the session setup in the event of a failure.

The main functional component in the I-CSCF originates a diameter Cx transaction to the HSS to obtain the routing information for the serving-CSCF. It therefore uses the

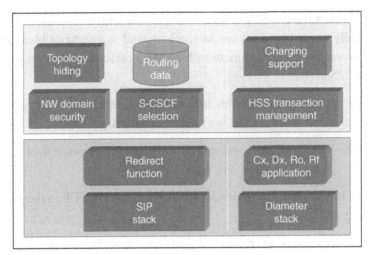

Figure 9.6 The I-CSCF components.

diameter framework for the transaction and response. Upon receiving the response, it redirects the transaction to the route of the appropriate S-CSCF.

Since the I-CSCF also functions as the entry point in a home network for a session that is originating from a visited network, it also needs to support network domain security for ensuring that the call is originating from a trusted network. It also validates the Visited-Network-ID header for validating the roaming agreement.

9.3.4.3 The S-CSCF

The S-CSCF plays a central role in the session plane to perform the following:

- It performs the leg management and call control function

- It enables, obtains, and stores the user profiles from the HSS, and communicates the registration information

- It obtains the authentication vectors from the HSS and enables the DIGEST authentication.

- It maintains a registrar function for the registered users and their corresponding routing information.

- It hosts the Initial Filter Criteria (iFC) processing for service invocation to the HSS.

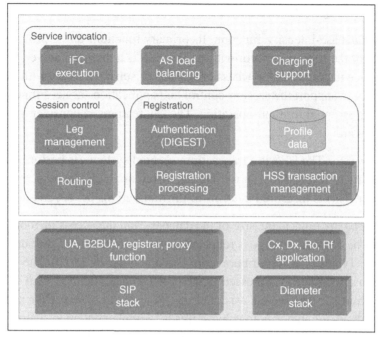

Figure 9.7 The S-CSCF components.

The S-CSCF function maintains state information as a stateful SIP server. It plays the role of a UA, B2BUA, registrar, and proxy server based on the services it supports. It therefore must be able to support a redundancy paradigm of $1 + 1$, so that a backup instance can continue to support the session setup in the event of a failure.

The main functional modules in the S-SCSF can be classified into three categories as seen in Figure 9.7. To support the Registration function, the functions obtain the user profile and authentication vectors from the HSS and store them locally in memory. These are requested with the Diameter Cx interface. The S-CSCF subsequently processes the registration requests and maintains a registry of the users and their routing information. Leg management is the central function of session control by which the S-CSCF can coordinate the inbound legs and outbound legs required for a session. This S-CSCF function has the ability to initiate and maintain legs to multiple application servers and media servers, in addition to the destination party. Finally, the service invocation function in the S-CSCF comprises the iFC function and the execution logic for the filter criteria. This function also extends a load balancing function to select an AS.

9.3.5 Implementing an HSS

The HSS is a database-intensive function. Its primary functions are to provide the access to and storage of the subscriber information. The HSS is a diameter server, which must be able to service the requests for access or updates of subscriber data with low response times. Its ability to support the session-control and application services functions in the IMS network is essential for their operation. The key features to consider in an HSS implementation are:

- *Performance* The HSS is a transaction-oriented system. Its performance reflects its ability to respond to a subscriber data fetch or an update request in a timely manner. This is measured in terms of the transactions per second it can process.

- The *Capacity* reflects the size of the database and the access methods to retrieve data. The database storage is reflected in terms of the size of the subscriber record. Typically, a subscriber record is in the order of 2.5 Kb. The large part of the subscriber data is expressed as the profile-related fields, which we examined in Section 5.2. The second aspect of the storage is to optimize the access methods. A suitable way to achieve this is to implement an in-memory cache mechanism and the persistent storage in a Redundant Array of Independent Disks (RAID) database configuration.

- *Scalability* is a careful consideration for the HSS. An HSS for a new service introduction or a small network, may constitute a few hundred thousand subscribers. However, as more subscribers are provisioned, the database can grow to support several million subscribers. However, not all the subscribers in a single large network need to be provisioned on the same HSS. These can be provisioned across multiple HSS systems.

- *Availability and serviceability* is a requirement for most core network elements, and refers to the ability for the HSS to support session control and other services within limits of an acceptable downtime. An availability of 99.999% or 5-nines, refers to being able to provide service with no more than 5 minutes of downtime in a year.

With these characteristics in perspective, an HSS can be constructed with a distributed model as shown below. This comprises a set of front-end processors (FEP) and a set of back-end processors (BEP) communicating via an Ethernet switching fabric. The basis

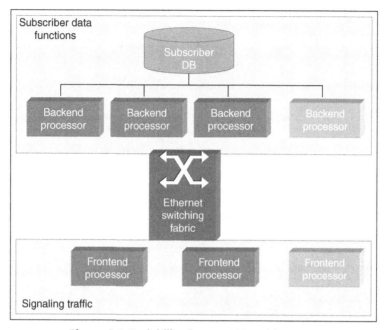

Figure 9.8 Scalability for an HSS architecture.

behind this architecture split is to partition the signaling and database functions across the two sets. Keeping them independent and modular also helps in scalability. The front-end processor set provides traffic scalability and the back-end processor set provides subscriber scalability.

The FEPs handle the signaling traffic, which are the diameter command requests that originate from the CSCFs and application servers. The FEPs work independently of each other, but to the network they present as a single element. In other words, a signaling request can be routed in a load-shared mode to the next available FEP. Adding FEPs to this set can help increase the traffic-handling capability of the system, thereby providing traffic scalability. The FEP set can function in an N + M redundancy paradigm to support high availability.

The BEPs host the subscriber database server applications. The database itself can be hosted on a RAID, and is accessible from any of these BEPs. This helps in scalability, since additional BEPs can be added to increase the subscriber capacity. The BEP set can also function in an N + M redundancy paradigm to support high availability.

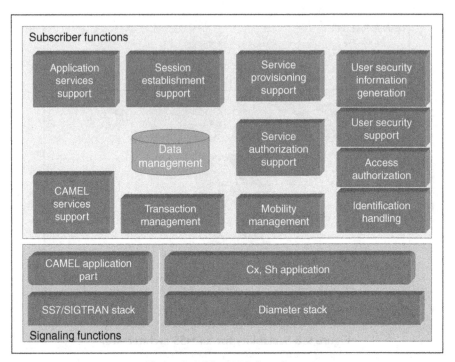

Figure 9.9 The HSS functions.

Figure 9.9 illustrates the details of the signaling and subscriber functions in the HSS. The signaling functions consist of the support for the diameter interfaces for the Cx and Sh applications for session control and the application services, respectively. To support CAMEL services, the HSS also requires an SS7/SIGTRAN stack with the CAMEL application part. The HSS provides the functions for enabling access authorization, subscriber identification, and generating authentication vectors. It supports the functions for session establishment and application services by making subscriber profile and location data available.

9.3.6 Implementing Media Plane Elements

The media plane as we recall, comprises two functions: the media resource function (MRF) and the media gateway function (MGF). These deliver services to the core network elements and the border internetworking elements, respectively. The MRF provides the media resources for the support of multimedia functions, which include transcoding, streaming, and conferencing. The MGF provides the media processing and media conversion between circuit-switched and packet-switched networks.

The media plane elements in IMS have been layered to maintain a logical and physical separation of media access and media control. This applies to both the MRF and the MGF. Both functions were recommended to use a common Megaco/H.248 model for control of the media access elements. While the practical realization of the MGF has followed this model, few vendors have supported it for the MRF. This is also a work item in 3GPP Release 7. The MRF has consequently emerged with two implementation models.

- *The MRFC/MRFP model with Megaco/H.248 control and SIP external interface.* The Media Resource Function Controller (MRFC) and the Media Resource Function Platform (MRFP), which provide these functions, logically communicate over the Mp interface. The Mp interface uses the Megaco/H.248 for the control of the MRFP resources. The MRF is also required support for the SIP-based Mr interface, so that the S-CSCF can request the media resources for streaming and other multimedia applications.

- *A client-server-based Media Server model.* The media server model is an evolution of the media function from the VoIP networks and the IETF standards. The premise for this model is that enhanced services work better with client-server models than the master/slave access model of Megaco/H.248. Several specialized application servers such as Interactive Voice Response (IVR), announcement servers, and so forth in implementation today, already use markup-based protocols to implement media control.

Megaco/H.248 has been well accepted for the Media Gateway Controller (MGC)-Media Gateway (MGW) model. The nature of the media access in the MGW is transport-oriented. The media streams need to be routed to their endpoints appropriately, with the correct encoding of the media formats. The MGF as it evolved from the VoIP networks, also played a more significant switching role. The MRF is not seen to require the master/slave features for connection control. Rather it is seen as a lighter function, which needs to provide the access methods for resource control. The next argument against Megaco/H.248 is that the access methods it exposes are too low level for application developers. Few applications or developers operate at the level of media resources as such (RTP streams, Dial Tone MultiFunction (DTMF) detectors, players, mixers, and so forth). This is acceptable for the low-level transport-oriented control of Digital Signal Processor (DSP) resources. Application server developers require access methods in terms of the call and leg control model, where a resource in terms of a media stream, announcement, or conference can be applied. Several markup-based methods have emerged to provide the right set of abstractions and the control mechanisms. These are aimed to provide a suitable alternative to Megaco/H.248 for use with media servers.

- *VoiceXML (VXML)* models the voice response or voice browsing function analogous to the HTML client-server model. It specifies the XML-based format to implement audio dialogs in a voice response application. The VXML scripts are provided to the media server, which interprets them as a "browser" application and provides the necessary resources to support the voice dialog.

- *Call Control XML (CCXML)* extends telephony call control support for VXML. The VXML scripts are limited in terms of supporting call and leg management functions. CCXML provides the XML support for the control, the setup, monitoring, and tear down of phone calls.

- *Network Announcement (netann)* is defined from the IETF Basic Network Media Services with SIP standard. This is a SIP-based client-server protocol to enable a media server to provide voice announcements. Netann has the ability to invoke VoiceXML scripts.

- *Media Sessions Markup Language/Media Objects Markup Language (MSML/MOML)* provide a SIP-based implementation for media server control. To a large extent they follow the transport-oriented paradigm of Megaco/H.248.

- *Media Server Command* Markup *Language (MSCML)* specified in RFC 4722, enables advanced conferencing capabilities.

- *Media Resource* Control *Protocol (MRCP)* specified in RFC4463, is a SIP-based protocol to utilize text-to-speech (TTS) and automatic speech-recognition (ASR) functions from VoiceXML.

9.3.6.1 MRFP

An MRFP provides the resources for both video and audio media processing. The resources for these functions are implemented in special purpose Digital Signal Processing (DSP) chips and hardware. These functions include codec, transcoding, record and playback, and other algorithms. Other mechanisms such as buffering to prevent jitter are also supported as a part of the hardware. The manipulation of and access to these resources is provided at a higher layer implemented as part of the host-based operating environment. These are exposed as APIs to an application layer. The MRFP application provides the interface capability to handle the message interaction with the MRFC. The message command processing may support the master/slave model or the SIP with markup-based client-server model.

9.4 Implementing Application Servers

IMS application servers perform specialized tasks. Based on the service logic they support, the application servers have a different set of performance requirements. These requirements typically may not be as stringent as the core network elements. IMS application servers also play a vital role in delivering converged multimedia services. Thus their ability to inter-work with other services such as Web-based, legacy network is essential as well.

There have been three approaches to realize application servers in the IMS network. The new generation of IP multimedia services can be built as SIP servers with the support of the necessary diameter applications. These can be built by developing the application logic on a SIP server framework foundation. The Java Platform Enterprise Edition (J2EE) specification provides a suitable framework for this set of applications. The second approach of extending legacy applications to IMS, can be done by either of two methods. The first method involves IMS-enabling existing legacy servers. In other words, it is to provide the necessary support for the SIP ISC and the Diameter Sh interfaces to an existing service platform. The second method involves building a bridging function or a gateway to access the legacy service. This requires the necessary inter-working between the SIP ISC and Diameter Sh to the interfaces of the Legacy server, such as CAMEL/IN and/or OSA/Parlay. The third approach addresses the problem of service interaction, and requires a platform to be able to interact between SIP application servers and/or legacy services. This can be realized by a combination of the previous two approaches, augmented with a rule-based service definition and interaction logic.

9.4.1 SIP Servers

HTTP services have been widely used in Web application development. Leveraging this model in IMS provides an effective bridge for converging with Web services, and also

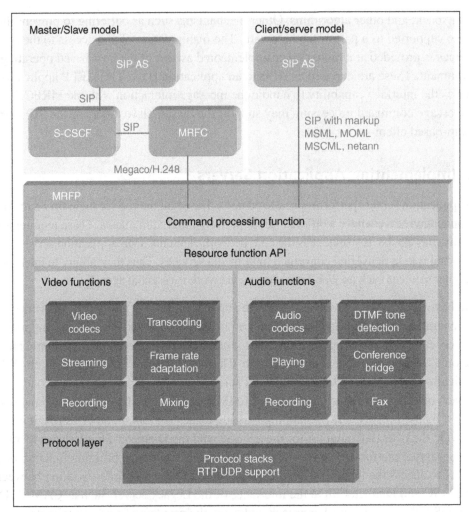

Figure 9.10 The MRF components.

opens up applications to a familiar development community. JSR116 extends the HTTP
service paradigm for SIP. HTTP services have evolved from monolithic Web server models
to lightweight servlets. A *servlet* is a Java program that runs as part of a network service
implemented as a server, and responds to client requests. Servlets execute within a *container*.
A container provides the low-level run-time support for the application components, and
also provides the necessary methods and protocols that are required by the application

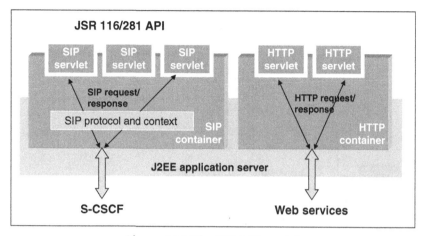

Figure 9.11 The JSR 116 model.

components through a Java API. A servlet container allows the ability to automatically translate specific resource locators (URL) into servlet requests. The servlet container is aware of the servlet capabilities and addressing. The container can initialize the servlets and deliver them the requests to execute. Servlets can be added or removed dynamically.

Since SIP call control is different from HTTP invocation, JSR 116 addresses the necessary modifications required for SIP servlets. The advantage is that services implemented as SIP servlets can interact with existing HTTP servlets to create converged applications. The JSR 116 model defines the SIP servlet API to execute within a servlet container. The container provides the SIP protocol stack. This aids in simplicity for the application to populate only the required SIP headers. Also, protocol aspects such as message retransmission and transaction management are done transparently of the application. The container also handles the context and state management of the SIP protocol. This enables the state complexity to be managed by the container and not by the servlet.

9.4.2 SCIM

The nature of the SCIM to support service interaction between SIP application servers, lends itself well to the servlet model. The servlets can be invoked through a rule-based engine, which defines the logic for service interaction. This can be made effective by cascaded execution of the servlets. JSR 289, which is still in a draft stage at the time of writing, extends the JSR116 model to support this feature.

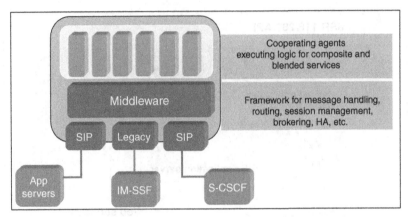

Figure 9.12 The SCIM in a converged environment.

The more challenging part is to gauge the interfaces to various application servers. The natural choice is to extend the IMS Service Control (ISC) interface towards the application servers as well, which provides a homogenous view to both SIP and legacy servers. However, an alternate model is to realize the value of the SCIM by providing service mediation between existing revenue-generating services and next-generation services. Figure 9.12 shows the layered model. The SCIM needs to be able to support the necessary interfaces and adapters to Legacy service elements (e.g., an IN SCP). The next layer comprises the middleware to support message handling, routing, and transaction management. The Execution layer based on either the servlet model or implemented as a set of cooperating agents directed by a rule engine, provides the support for creating a composite service between the various interfaces.

9.5 Managing the IMS Network

Having explored the various methods to implement IMS network elements, we now conclude with the IMS network itself. It is essential to integrate IMS into the service framework of the MNO/service provider, so that it can be effectively managed and operated to deliver the services it is intended for. The principles of managing the IMS network are similar to 2G and 3G systems. IMS is the core network for the 3G (and potentially the LTE and 4G systems). It therefore must follow uniform management as the other subsystems. The basic management principles for the IMS network are to provide:

- Inter-working with the 3G (3GPP/3GPP2/NGN) management systems.

- Standard communication between the various Network Elements (NEs) such as the CSCFs, HSSs, MRF and the Operations Systems (OS).

- A uniform method to manage multi-vendor NEs and their management systems.

- Configuration of the network elements, which includes specific parameters and bulk configuration.

- Fault management capabilities.

- Interfaces for exchange of charging and accounting information between different service providers.

- Performance indicators for managing the operation of the network.

- A scalable method to grow the network by adding the resources in the network as necessary.

- Remote management operations to minimize onsite support.

- System and software upgrades and backup and restore operations to ensure serviceability.

The ITU standard Telecom Management Network (TMN) M-series provides the architecture and interface definition for managing a diverse set of telecom networks. It addresses the relationships between the OS and the NEs, and defines the Q-interfaces between a single operator and the X-interfaces between different operators. The Telecom Operations Map (TOM) from the TeleManagement Forum, which we observe in Figure 9.13, extends the TMN to address customer-specific operations support and management. 3GPP and TISPAN recommend the management framework based on these two standards.

The TMN/TMF-based network architecture shown in Figure 9.14, adopts a top-down approach based on deriving the network and service management on business needs. At the top level there are specialized OS functions for customer interface management, customer care, and the network and systems management. The network and service management processes encompass the management of different domains, which include the user equipment, access network, and the core network. The interface between these

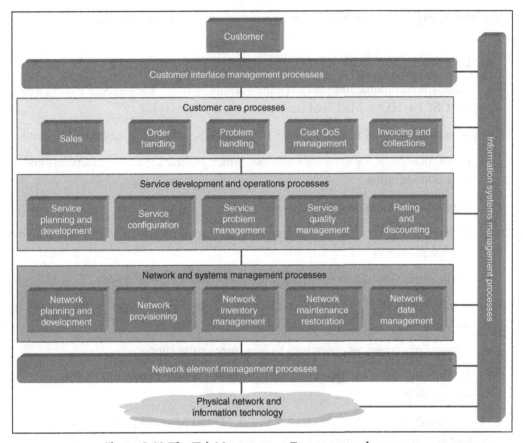

Figure 9.13 The TeleManagement Forum operations map.

OS functions is the Itf-N interface. IMS is a part of the core network domain. The management framework in the IMS domain follows a Manager-Element model, where the Element Manager (EM) supports the management processes of a group of the NEs. The interface between the EM and NE is also referred to as the Type 1 interface. The following Type 1 interfaces are defined for the exchange management information. The networking protocols are:

- Simple Network Management Protocol (SNMP)

- Common Object Request Broker Architecture Internet Inter-Orb Protocol (CORBA IIOP)

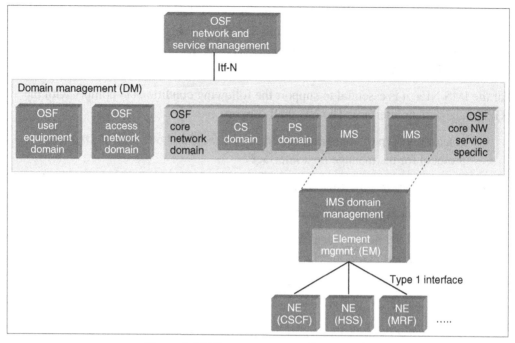

Figure 9.14 The management framework.

- Simple Object Access Protocol (SOAP)

- XML

The following file transfer methods are identified for the transfer of bulk information including charging and accounting data:

- File Transfer Access Method (FTAM)

- FTP, trivial FTP (TFTP), secure FTP (SFTP)

While IP is the transport for IMS network elements, most implementers separate the management interface from the signaling and traffic interface. In other words, the signaling for these protocols is assigned on a physically separate IP network interface.

3GPP guides the specification of the management interface for the NEs in terms of an Integration Reference Point (IRP). The IRP is defined in terms of the tasks specified in the TMF operations map. These requirements are specified as a technology independent information service. This is mapped to the protocol interfaces SNMP, CORBA, XML, and SOAP, which are specified as a Solution Set (SS), respectively. The Integration reference points are further categorized into specifying the interface, network resource model (NRM), and the data definition.

For the IMS NEs, it is essential to support the following conditions to comply with the 3GPP management interface. The NE shall support:

- at least one Type 1 standard networking protocol and one bulk file transfer protocol for the management interface

- a Q-interface adapter function, if it is unable to support the management interface

- the IRP solution sets related to the application protocols

- implement the management functionality for that information service specified by the relevant 3GPP interface specification

The management processes that utilize this framework and interfaces, comprise the well-known FCAPS and other functional areas that follow.

Fault management processes are responsible for the detection, isolation, reporting, and repair of faults. This implies that the IMS NEs must be able to trigger this function on the occurrence of a failure or fault condition. Fault reporting is done through alarms and event reports.

Configuration management (CM) focuses on the operational parameters and relationships for both network elements and network resources. Commonly used configuration data examples are the addressing, routing setup. The configurable elements are modeled as managed objects and their instances. These are organized in a hierarchical relationship also referred to as a management information base. Configuration data can be applied from

Table 9.1

FCAPS	
• Fault Management	• Subscription Management
• Configuration Management	• Software Management
• Accounting Management	• QoS Management
• Performance Management	• User Equipment Management
• Security Management	• Roaming Management
	• Fraud Management

a single management entity and in order to stage the data in the field, it can be phased as the active CM and passive CM. The ability to apply or retrieve configuration data in bulk from a single/multiple NEs through a central OS/EM is referred to as the Bulk CM.

Accounting management covers the principles of the online and offline charging of events, including the generation of the CDRs, which we examined in Chapter 5.

Performance management is based on the measurements and collection of the necessary data, which is essential to understand the operation of the system and compare it with the engineered goals. In Appendix A, we look at some of the performance criteria of the IMS network.

Security management covers the areas of both access and network domain security models, which we explored in Chapter 5. In addition, it also provides the access control mechanisms to users of the NEs, which are provided in terms of authorization to perform certain tasks at specific times.

Subscription management is concerned with the provisioning of the subscriber and service profile in the HSS, and the appropriate configuration of the USIM.

Software management covers the delivery, install, and activation of the software for the IMS NEs. This covers provisioning the software for a first time install and also upgrades in a live network.

Roaming management covers the necessary contractual and usage agreements with other service providers, to accommodate the roaming of a customer from one's home network to a visited network. *Fraud management* provides the functions to detect and prevent fraud, especially with roaming and mobility services.

QoS management consists of two areas: provisioning the QoS policy and monitoring the QoS. The QoS policy provisioning is essential in IMS, to ensure the end-to-end service quality for the user applications. The provisioning is done essentially in the PCRF, AF and AGW functions. The QoS monitoring is done by the collection of data across all the network elements. This includes the alarm event reports and the performance measurement data.

9.6 Chapter Summary

The implementation choices for the functional elements in the IMS architecture are realized by industry-driven approaches. The good part is that equipment vendors are moving towards standards-based approaches from proprietary implementations, thereby creating an ecosystem that can be built with a multi-vendor solution. Implementation standards have evolved to help converge with other network solutions as well. Java-based specifications for the handset and the application servers is an example. The JSR-281 and JSR-116/289 specifications are helping to provide the foundation for the IMS handset client applications and the application servers. These align with the existing base of mobile handset environment and the Web-based application servers, respectively. Media elements that have been maturing from the VoIP implementations, are offering alternate approaches to use a client-server control model than a megaco/H.248 master/slave recommendation.

The core network elements on the other hand, are performance-intensive. The need to minimize latency in the path of the session establishment and control requires developers to focus on efficient code and maximizing the capabilities of the hardware. Blade server-based hardware architecture is providing the base for performance, reliability, and scalability for these elements.

As a network, IMS leverages the emerging IP/MPLS backbone, which is gaining acceptance in the service provider network. Managing the IMS network also requires for it to fold into the management framework of the 3G network and coexist with the management subsystems.

CHAPTER 10

Epilogue: Future Directions for IMS

IMS is a journey not a destination

It's still a way to go to completion

Watch the beauty of its evolution

There is more to come from innovation

We embarked on our journey to understand IMS with the definition as it stands with Release 7 of the 3GPP standards. The evolution continues as 3GPP progresses with the work items in Release 8. Let's take a peek at some of the areas that are currently being worked on that will help shape newer features in this architecture. Release 7 work areas are continuing to improve features including Voice Call Continuity, CS-switched IMS combinational services such as VideoSharing, Emergency Calling, Policy Control and Charging, End-to-end QoS, Telecoms & Internet Converged Services & Protocols for Advanced Networks (TISPAN), and PacketCable features integration, and 3GPP Security Inter-working. The other new areas to integrate are:

- Multimedia Telephony (MMTEL)

- Service Identification using IMS Communication Service Identifier (ICSI)/IMS Application Reference Identifier (IARI)

- Parlay X WS extensions to support Geocoding, Application-driven QoS, multimedia streaming, and multicasting control.

- ISIM API for Java Card

- Enhanced Universal Subscriber Identity Module (USIM) Phonebook

- Lawful Intercept support

The following are Release 8 work areas:

- Service Broker

- Multimedia Priority Service

- Support for machine-to-machine communication

- Operational, Administration, and Maintenance Procedures Support

- Personal Network Management (PNM)

10.1 Innovation Beyond Standards

Research and development has been difficult with the traditional telecommunication systems, due to their limited accessibility. The OpenSource IMS (OSIMS) Core Project aims to change that for the IMS networks. It is aiming to provide an IMS core reference implementation. This benefits the prototype IMS applications and test IMS as a technology in a research environment. The OSIMS supports an implementation of the CSCFs and a lightweight Home Subscriber Server (HSS). These are based on the open source SIP Express Router.

Innovation has also made the concept of IMS in-a-box closer to reality. Traditional Core telecom network infrastructure equipment could be delivered in a large volume of server racks, most of which was occupied by Time Division Multiplexing (TDM) switching. The IP core network is transforming that. With the blade server-based approaches we saw in the previous chapter, the concept of IMS in a single chassis has been demonstrated as a proof-of-concept. With each blade server hosting a functional element of the IMS core network such as the S-CSCF, P-CSCF, I-CSCF, HSS, AS, MRF, and so forth, a single 20-inch server can potentially deliver a sizeable functionality of the IMS architecture.

10.2 The Promise and the Wait

While work on the IMS standards started circa 2002, it started generating a major industry buzz around 2005. Its promises, some of which we have seen in this book, set some very high expectations for the adoption of this architecture. Operators have continued to trial IMS applications and have deployed point applications. The wait to see this go mainstream has unnerved the impatient as well. IMS continues to follow the trajectory of the 3G systems. While there have been interim solutions for fixed mobile convergence, multimedia applications such as video on high-speed 2.5G networks, these are not disruptive to IMS. There is still untapped potential in 2.5G networks, which has been shown by the iPhone and the Blackberry. Upon its saturation, the 3G networks will get an impetus and IMS will get a greater thrust.

While IMS by itself may seem a complex architecture, its availability makes some of the next generation access networks such as femtocell and mobile broadband elements much simpler. IMS helps promote the Flat IP architecture, which simplifies these emerging network elements. The demand for mobile broadband, mobile multimedia, and Web 2.0 applications, and wider adoption of 3G are the accelerating factors for IMS.

On a final note, everyone wants to know what the "killer app" is for IMS, if there is one. Clearly, multimedia services will transform the means of our communication and a blending of these applications will deliver the killer app for IMS. It would be speculation to point to any of the applications we saw in this book, but these will pave the way for an exciting new world of communication.

I hope you found this book informative. I have tried my best to capture content both as it is written in the standards and interpreted by the industry. This is a vast and evolving topic and I hope that any errors and omissions are accepted. I would welcome your comments, suggestions, and criticism at ahanda@computer.org.

Arun Handa

Melbourne, FL

March 2008

Performance of IMS Networks

The performance of the IMS network is important to understand to apply it in effective network operation and network planning. Measurement data from the network elements, which is collected for the engineered traffic and QoS, gives an insight into how well the system is performing relative to the desired goals. These performance measurements are specified in 3GPP TS 32.409, which define the counters and data for the various procedures and events at the network elements. These are summarized in Table A.1. Most measurement counters provide qualified data about the conditions that apply, such as successful and failure conditions.

Table A.1 Performance measurements for the IMS.

NE	Measurement (Counters and Data)	Qualifiers
CSCF	Initial registration procedures	Attempted, Successful, Failed
	Initial registration procedures per access technology	Attempted, Successful, Failed
	Mean Initial registration setup time	
	Re-registration procedures	Attempted, Successful, Failed
	Re-registration procedures per access technology	Attempted, Successful, Failed
	De-registration procedures per access technology	Attempted, Successful, Failed
	HSS-initiated De-registrations	

Continued

Table A.1 (continued)

NE	Measurement (Counters and Data)	Qualifiers
	S-CSCF-initiated De-registrations	
	PSI registrations	Attempted, Successful, Failed
	S-CSCF registration/deregistration notification procedures	Attempted, Successful, Failed
	Session establishments	Attempted, Successful, Failed
	Simultaneous online and answered sessions (Maximum)	
	User location query procedures	Attempted, Successful, Failed
	Attempted session establishments from users of other network domains	
	403 (Forbidden) generated for sessions from users of other network domains	
	Attempted session establishments to users of other network domains	
	403 (Forbidden) received for sessions to users of other network domains	
	Initial registrations of visiting users from other IMS network domains	
	Number of SIP 403 (Forbidden) messages for the roamers	
	Roaming users to other IMS network domains	
	Authentication procedures	Attempted, Successful, Failed
	HSS initiated User Profile Update procedures	Attempted, Successful, Failed
	Subscription procedures	Attempted, Successful, Failed
	Notify procedures	Attempted, Successful, Failed
HSS	Number of provisioned IMS subscriptions currently stored in a HSS	

Table A.1 (continued)

NE	Measurement (Counters and Data)	Qualifiers
	Number of provisioned Private User Identity stored in an HSS	
	Number of provisioned Public User Identities with SIP URI format stored in an HSS	
	Number of provisioned Public User Identities with TEL URI format stored in an HSS	
	Number of Private Service Identities stored in an HSS	
	Number of Public Service Identities with SIP URI format stored in an HSS	
	Number of Public Service Identities with TEL URI format stored in an HSS	
	Registered Public User Identities	
	Un-Registered Public User Identities	
	Registered Private User Identities	
	User registration status queries	Attempted, Successful, Failed
	S-CSCF registration/de-registration notifications	Attempted, Successful, Failed
	Network initiated de-registrations by HSS	Attempted, Successful, Failed
	User location queries	Attempted, Successful, Failed
	Authentication procedures	Attempted, Successful, Failed
	Number of UDR procedures	Attempted, Successful, Failed
	Number of PUR procedures	Attempted, Successful, Failed
	Number of SNR procedures	Attempted, Successful, Failed
	Number of PNR procedures	Attempted, Successful, Failed
BGCF	Sessions at BGCF	
	Sessions forwarded to MGCF	Attempted, Successful, Failed
	Sessions forwarded to BGCF	Attempted, Successful, Failed

Continued

Table A.1 (continued)

NE	Measurement (Counters and Data)	Qualifiers
MGCF	CS network originated calls	Attempted, Successful, Answered, Failed
	Call setup time (Mean), CS network originated	
	IM CN originated calls	Attempted, Successful, Answered, Failed
	Call set-up time (Mean), IM CN originated	
	Call release initiated by CS network	
	Call release initiated by IM CN subsystem	
	Call release initiated by MGCF	
	Call release initiated by MGW	
	Number of simultaneous online and answered calls (Maximum)	
	Number of simultaneous online and answered calls (mean)	
MRFC	Session establishments	Attempted, Successful, Failed
	Multi-party session establishments	Attempted, Successful, Failed
	Event subscription procedures for multi-party sessions	Attempted, Successful, Failed
	Session establishments for announcements	Attempted, Successful, Failed
	Session establishments for transcoding service	Attempted, Successful, Failed
MRFP	Number of RTP packets	Incoming, Outgoing, Lost
	Size (Octet) of RTP packets	Incoming, Outgoing
PCRF	Resource authorization procedures at session establishment	Attempted, Successful, Failed
	Resource authorization procedures at session modification	Attempted, Successful, Failed
	Resource reservation procedures	Attempted, Successful

Table A.1 (continued)

NE	Measurement (Counters and Data)	Qualifiers
	Failed resource reservation procedures by PCRF	
	Failed resource reservation procedures by AGW/GGSN	
	Authorization of PDP context modification	Attempted, Successful, Failed
	Indication of PDP context modification	
SLF	Routing information interrogation procedures	Attempted, Successful
	Routing information interrogation procedures from CSCF	Attempted, Successful
	Routing information interrogation procedures from AS	Attempted, Successful
AS	IM CN originated session establishments	Attempted, Successful
	AS originated session establishments	Attempted, Successful
	Attempted incoming messages, Page-Mode	Attempted, Successful
	Attempted incoming messages, Session-Mode	Attempted, Successful
	Conference creations	Attempted, Successful
	Conference joinings	Attempted, Successful
	Conference invitations	Attempted, Successful
	Conference subscriptions	Attempted, Successful
	Number of simultaneous online users (Maximum)	
	Number of simultaneous online conferences (Maximum)	
	Number of simultaneous online users (Mean)	
	Number of simultaneous online conferences (Mean)	

Continued

Table A.1 (continued)

NE	Measurement (Counters and Data)	Qualifiers
	Number of simultaneous online watchers (Maximum)	
	Number of simultaneous online presentities (Maximum)	
	Number of simultaneous online watchers (Mean)	
	Number of simultaneous online presentities (Mean)	
	Presence subscriptions	Attempted, Successful
	Presence notifications	Attempted, Successful
	Presence publications	Attempted, Successful
	PoC session Creation	Attempted, Successful
	PoC session Joining	Attempted, Successful
	PoC session Invitations	Attempted, Successful
	Number of simultaneous online users (Maximum)	
	Number of simultaneous online PoC session (Maximum)	
XDM Enabler	HTTP Put procedures	Attempted, Successful
	HTTP Get procedures	Attempted, Successful
	HTTP Delete procedures	Attempted, Successful
	Subscription procedures	Attempted, Successful
	Notification procedures	Attempted, Successful
Equipment	Mean processor usage	
	Peak processor usage	
	Mean memory usage	
	Peak memory usage	

Glossary

3GPP	3rd Generation Partnership Project
3GPP2	3rd Generation Partnership Project (2)
AAA	Authentication, Authorization, Accounting
ABMF	Account Balance Management Function
AGW	Access Gateway
AH	Authentication Header
AJAX	Asynchronous JavaScript and XML
AKA	Authentication and Key Agreement
ALG	Application Layer Gateway
ALI	Automatic Location Information
AMR	Adaptive Multirate
ANI	Application Network Interface
	Automatic Number Identification
ANSI	American National Standards Institute
AOR	Address of Record
APN	Access Point Name
A-RACS	Access Resource and Admission Control Subsystem
AS	Application Server
ATCA	Advanced telecommunications Communications Architecture
AVP	Attribute Value Pair
B2BUA	Back to Back User Agent
BC	BroadCast
BGCF	Border Gateway Control Function
BICC	Bearer Independent Call Control
BSS	Business Support System
CAMEL	Customized Applications for Mobile networks Enhanced Logic
CCF	Charging Control Function
CDF	Charging Data Function
CDMA	Code Division Multiple Access

CDR	Call Data Record
CMTS	Cable Modem Termination System
CN	Core Network
COD	Content on Demand
COPS	Common Open Policy Service
CS	Circuit Switched
CSAF	Circuit Switched Adaptation Function
CSRN	Circuit Switched Routing Function
CTF	Charging Trigger Function
DCCA	Diameter Credit Control Application
DHCP	Dynamic Host Configuration Protocol
DNS	Domain Naming Service
DSLAM	Digital Subscriber Line Access Multiplexer
DSP	Digital Signal Processer
DTF	Domain Transfer Function
DWR	Direct Web Remoting
EAP	Extensible Authentication Protocol
ECF	Event Charging Function
EDGE	Enhanced Data Rates for GSM Evolution
ESP	Encapsulated Security Payload
ETSI	European Telecommunication Standards Institute
EVDO	Evolution Data Optimized
FBC	Flow Based Charging
FMC	Fixed Mobile Convergence
GGSN	Gateway GPRS Serving Node
GPRS	General Packet Radio Service
GRUU	Globally Routable User agent Uniform resource identifier
HLR	Home Location Register
HSS	Home Subscriber Server
HTTP	HyperText Transfer Protocol
IBCF	Interconnect Border Control Function
ICID	IMS Charging Identifier

I-CSCF	Interrogating Call Session Control Function
IE	Information Element
IETF	Internet Engineering Task Force
IFC	Initial Filter Criteria
IKE	Internet Key Exchange
IMPI	IMS Private Identity
IMPU	IMS Public Identity
IMRN	IMS Routing Number
IMS	IP Multimedia Subsystem
IMSI	International Mobile Subscriber Identity
IM-SSF	IMS Service Switching Function
IN	Intelligent Network
IP	Internet Protocol
IP-CAN	IP Connectivity Access Network
IPSec	IP Security
IPTV	Internet Protocol Television
IPv4	Internet Protocol version 4
IPv6	Internet Protocol version 6
ISC	IMS Service Control
ISIM	IMS Subscriber Identity Module
ISP	Internet Service Provider
ISUP	ISDN User Part
ITU	International Telecommunications Union
JSR	Java Specification Request
LRF	Location Resource Function
LSM	Leg State Model
LTE	Long Term Evolution
MDN	Mobile Directory Number
MEGACO	**Me**dia **Ga**teway **Co**ntrol Protocol
MGCF	Media Gateway Control Function
MGF	Media Gateway Function
MMD	Multimedia Domain
MMS	Multimedia Messaging Service

MNO	Mobile Network Operator
MRF	Media Resource Function
MRFC	Media Resource Function Controller
MRFP	Media Resource Function Platform
MSAG	Master Street Address Guide
MSC	Mobile Switching Center [V-visited G-Gateway]
MSISDN	Mobile Subscriber ISDN Directory Number
MSRP	Message Session Relay Protocol
MTU	Message Transmission Unit
NA(P)T	Network Address Port Translation
NAI	Network Address Indicator
NASS	Network Attachment SubSystem
NAT	Network Address Translation
NNI	Network Network Interface
N-PVR	Network Personal Video Recorder
OCF	Online Charging Function
OMA	Open Mobile Alliance
OSA SCS	Open Service Access Service Capability Server
OSI	Open Systems Interconnect
OSS	Operational Support System
PAM	Packet Cable Access Manager
PCC	Policy Charging and Control
PCEF	Policy Charging Enabling Function
PCRF	Policy Charging and Resource Function
P-CSCF	Proxy Call Session Control Function
PDG	Packet Data Gateway
PDP	Packet Data Protocol
PDSN	Packet Data Serving Node
PEA	Presence External Agent
PES	PSTN Emulation Subsystem
PNA	Presence Network Agent
PoC	Push to Talk over Cellular
PS	Packet Switched

PSAP	Public Safety Access Point
PSI	Public Service Identity
PSTN	Public Services Telephone Network
PUA	Presence User Agent
QoE	Quality of Experience
QoS	Quality of Service
RACS	Resource Admission and Control Subsystem
RAN	Radio Access Network
RFC	Request For Comments
RSVP	Reservation Protocol
RTCP	Real Time Control Protocol
RTP	Real Time Protocol
RTSP	Real Time Streaming Protocol
SA	Security Association
SAD	Security Association Database
SAE	System Architecture Evolution
SBC	Session Border Controller
SBLP	Service Based Local Policy
SCIM	Service Capability Interaction Manager
SCP/SCF	Service Control Point/Service Control Function
S-CSCF	Serving Call Session Control Function
SCTP	Stream Control Transmission Protocol
SDP	Session Description Protocol
SDP	Service Delivery Platform
SDP	Session Description Protocol
SEG	Security Gateway
SGW	Signaling Gateway
SIM	Subscriber Identity Module
SIP	Session Initiation Protocol
SLF	Subscriber Locator Function
SMS	Short Messaging Service
SMSC	Short Messaging Service Center
SOA	Service Oriented Architecture

SOAP	Simple Object Access Protocol
SPD	Security Policy Database
SPDF	Service Policy Decision Function
SPT	Service Point Trigger
SSP/SSF	Service Switching Point/Service Switching Function
TCP	Transmission Control Protocol
TDM	Time Division Multiplexing
TISPAN-NGN	Telecoms & Internet converged Services & Protocols for Advanced Networks Next Generation Network
TLS	Transport Layer Security
TMN	Telecommunications Management Network
TrGW	Trunking Gateway
TTL	Time To Live
UA	User Agent
UAS/UAC	User Agent Server / User Agent Client
UDDI	Universal Description, Discovery and Integration
UDP	User Datagram Protocol
UDVM	Universal Decompressor Virtual Machine
UE	User Equipment
UICC	Universal Integrated Circuit Card
UNI	User Network Interface
URL	Uniform Resource Locator
USIM	Universal Subscriber Identity Module
VCC	Voice Call Continuity
VDI	VCC Domain Identifier
VDN	VCC Domain Number
VoD	Video on Demand
VoIP	Voice over IP
WLAN	(Public) Wireless Local Area Network
WSDL	Web Services Definition Language
XCAP	XML Capabilities Application Protocol
XML	Extensible Markup Language

References and Further Reading

3GPP/3GPP2 Standards

The 3GPP standards are written as Technical Specifications (TS) and Technical Reports (TR). These can be accessed from the 3GPP website at http://www.3gpp.org/ftp/Specs/html-info/*nn*-series.htm. *nn* is the number of the TS series. Most IMS standards fall between the 22-series and 33-series.

General

3GPP TS 22.101	Service aspects; Service principles
3GPP TR 22.975	Service aspects; Advanced Addressing
3GPP TS 22.228	Service requirements for the Internet Protocol (IP) multimedia core network subsystem; Stage 1
3GPP TS 23.002	Network architecture
3GPP TS 23.003	Numbering, addressing and identification
3GPP TS 23.221	Architectural Requirements

IMS Principles and Session Control

3GPP TS 23.218	IP Multimedia (IM) session handling; IM call model; Stage 2
3GPP TS 23.228	IP Multimedia (IM) subsystem; Stage 2
3GPP TS 24.228	Signaling flows for the IP multimedia call Control based on SIP and SDP; Stage 3

3GPP TS 24.229	IP Multimedia Call control based on SIP and SDP; Stage 3
3GPP TR 24.930	Signaling flows for the session setup in the IP Multimedia core network Subsystem (IMS) based on Session Initiation Protocol (SIP) and Session Description Protocol (SDP); Stage 3

Media

3GPP TS 26.114	IP Multimedia Subsystem (IMS); Multimedia Telephony; Media handling and interaction
3GPP TS 26.141	IP Multimedia System (IMS) Messaging and Presence; Media formats and codecs
3GPP TS 26.235	Packet-switched Multimedia applications; Default Codecs
3GPP TS 26.071	AMR speech Codec; General description
3GPP TS 29.332	Media Gateway Control Function (MGCF) – IM Media Gateway (IM-MGW); Mn interface
3GPP TS 29.333	Multimedia Resource Function Controller (MRFC) – Multimedia Resource Function Processor (MRFP) Mp interface;
3GPP TS 23.333	Multimedia Resource Function Controller (MRFC) – Multimedia Resource Function Processor (MRFP) Mp interface: Procedures Descriptions
3GPP TS 29.163	Interworking between the IP Multimedia (IM) Core Network (CN) subsystem and Circuit Switched (CS) networks

Diameter and Applications

3GPP TS 29.228	IP Multimedia (IM) Subsystem Cx and Dx interfaces; Signaling flows and message contents
3GPP TS 29.229	Cx and Dx interfaces based on the Diameter protocol; Protocol details
3GPP TS 29.328	IP Multimedia (IM) Subsystem Sh interface; Signaling flows and message contents
3GPP TS 29.329	Sh interface based on the Diameter protocol; Protocol details
3GPP TS 29.210	Charging rule provisioning over Gx interface
3GPP TS 29.211	Rx Interface and Rx/Gx signaling flows
3GPP TS 32.299	Diameter Charging Applications *(for Rf/Ro interfaces)*

Messaging

3GPP TR 22.940	IP Multimedia Subsystem (IMS) messaging
3GPP TS 22.340	IP Multimedia Subsystem (IMS) messaging; Stage 1
3GPP TS 24.247	Messaging Service using the IP Multimedia (IM) Core Network(CN) subsystem; Stage 3
3GPP TR 23.804	Support of SMS and MMS over generic 3GPP IP access

Group management, Presence, PoC and Conferencing

3GPP TS 22.250	IP Multimedia Subsystem (IMS) Group Management; Stage 1
3GPP TS 22.141	Presence service; Stage 1
3GPP TS 23.141	Presence service; Architecture and functional description; Stage 2
3GPP TS 24.141	Presence service using the IP Multimedia (IM) Core Network (CN) subsystem; Stage 3
3GPP TS 33.141	Presence service – Security
3GPP TS 23.979	Push-to-talk over Cellular (PoC) services; Stage 2
3GPP TS 24.147	Conferencing using the IP Multimedia (IM) Core Network (CN) subsystem; Stage 3
3GPP TR 22.948	Study of requirements of IP-Multimedia Subsystem (IMS) convergent multimedia conferencing

Combinational Services and Multimedia Telephony

3GPP TS 22.279	Combined Circuit Switched (CS) and IP Multimedia Subsystem (IMS) sessions; Stage 1
3GPP TS 23.279	Combining Circuit Switched (CS) and IP Multimedia Subsystem (IMS) services; Stage 2
3GPP TS 23.279	Combining Circuit Switched (CS) and IP Multimedia Subsystem (IMS) services; Stage 3
3GPP TS 22.173	IMS Multimedia Telephony Service and supplementary services; Stage 1
3GPP TS 24.173	IMS Multimedia telephony service and supplementary services; Stage 3
3GPP TS 24.247	Multimedia Telephony; Media handling and interaction

VCC

3GPP TS 23.206	Voice Call Continuity (VCC) between Circuit Switched (CS) and IP Multimedia Subsystem (IMS); Stage 2
3GPP TS 24.206	Voice Call Continuity between the Circuit-Switched (CS) domain and the IP Multimedia Core Network (CN) (IMS) subsystem; Stage 3
3GPP TR 23.806	Voice Call Continuity between CS and IMS Study

Policy and QoS

3GPP TS 23.107	Quality of Service concept and architecture
3GPP TS 23.207	End-to-end QoS concept and architecture
3GPP TR 23.208	End-to-end QoS Signaling Flows
3GPP TR 23.803	Evolution of policy control and charging
3GPP TS 29.214	Policy and charging control over Rx reference point

Charging

3GPP TS 22.115	Service Aspects – Charging and Billing
3GPP TS 23.125	Overall high-level functionality and architecture impacts of flow-based charging
3GPP TS 23.815	Charging implications of IMS architecture
3GPP TS 32.240	Telecommunication management; Charging management; Charging architecture and principles
3GPP TS 32.260	Telecommunication management; Charging management; IP Multimedia Subsystem (IMS) charging
3GPP TS 32.295	Charging management; charging data record (CDR) transfer
3GPP TS 32.296	Online Charging System (OCS); Applications and Interfaces

Security

3GPP TS 33.102	3G Security; Security architecture
3GPP TS 33.120	Security objectives and principles

3GPP TS 33.203	3G Security; Access Security for IP based services
3GPP TS 33.210	3G Security; Network Domain Security; IP network layer security
3GPP TS 33.828	IMS media plane security
3GPP TS 33.978	Security aspects of early IP Multimedia Subsystem (IMS)

Emergency Services, Location and CAMEL

3GPP TS 23.167	IP Multimedia Subsystem (IMS) emergency sessions
3GPP TR 23.867	Internet Protocol (IP) based IP Multimedia Subsystem (IMS) emergency sessions
3GPP TS 23.271	Functional stage 2 description of Location Services (LCS)
3GPP TS 23.278	Customized Applications for Mobile network Enhanced Logic (CAMEL); IP Multimedia System (IMS) interworking; Stage 2

USIM/ISIM

3GPP TS 21.111	USIM and IC card requirements
3GPP TS 31.102	Characteristics of the USIM application
3GPP TS 31.103	Characteristics of the IP Multimedia Services Identity Module (ISIM) application

Network Management

3GPP TS 32.101	Telecommunication management; Principles and high level requirements
3GPP TS 32.102	Telecommunication management; Architecture
3GPP TS 32.409	Performance Management (PM); Performance measurements – IP Multimedia Subsystem (IMS);
3GPP TS 32.600	Configuration Management (CM); Concept and high-level requirements;
3GPP TS 32.731	Telecommunication management; IP Multimedia Subsystem (IMS) Network Resource Model (NRM) Integration Reference Point (IRP): Requirements

3GPP2 Standards

The 3GPP2 website at http://www.3gpp2.org/Public_html/specs/index.cfm hosts the IMS standards for the North American and Asian ANSI/TIA/EIA-41 networks.

S.R0058-0	IP Multimedia Domain – System Requirements
X.S0013-000-B	Multi-Media Domain Overview
X.S0013-002-B	IP Multimedia Subsystem (IMS); Stage 2
X.S0013-003-B	IP Multimedia (IM) session handling; IM call model
X.S0013-004-B	IP Multimedia Call Control Protocol based on SIP and SDP; Stage 3
X.S0013-005-B	IP Multimedia (IM) Subsystem Cx Interface; Signaling flows and message contents
X.S0013-006-B	Cx Interface based on the Diameter protocol; Protocol details
X.S0013-007-A	IP Multimedia Subsystem; Charging Architecture
X.S0013-008-A	IP Multimedia Subsystem; Accounting Information Flows and Protocol
X.S0013-009-A	IMS/MMD Call Flow Examples
X.S0013-010-B	IP Multimedia Subsystem (IMS) Sh Interface signaling flows and message contents
X.S0013-011-B	Sh interface based on the Diameter protocol
X.S0013-012	Service Based Bearer Control – Stage 2
X.S0013-013	Service Based Bearer Control – Tx Interface Stage 3
X.S0013-014	Service Based Bearer Control – Ty Interface Stage 3
S.R0086-B	IMS Security Framework
S.R0095-0 v1.0	Support for IP Multimedia Services Identity Module (ISIM) on Universal Integrated Circuit Card (UICC) in 3GPP2 Systems – Stage 1 Requirements

TISPAN NGN Standards

The details about the TISPAN specifications are available at http://www.etsi.org/tispan/

The specifications can be obtained from http://pda.etsi.org/pda/queryform.asp

ETSI TR 180 000	NGN Terminology
ETSI TR 180 001	NGN Release 1; Release definition
ETSI TR 183 013	Analysis of relevant 3GPP IMS specifications for use in TISPAN_NGN Release 1 specifications
ETSI ES 282 001	NGN Functional Architecture Release 1 Overall architecture
ETSI ES 282 007	IP Multimedia Subsystem (IMS); Functional architecture
ETSI TS 182 006	IP Multimedia Subsystem (IMS); Stage 2 description IMS stage 2 Endorsement
ETSI ES 283 027	Endorsement of the SIP-ISUP Interworking between the IP Multimedia (IM) Core Network (CN) subsystem and Circuit Switched (CS) networks
ETSI ES 282 010	Charging [Endorsement of 3GPP TS]
ETSI ES 283 003	IP Multimedia Call Control Protocol based on Session Initiation Protocol (SIP) and Session Description Protocol (SDP) Stage 3
ETSI TR 183 032	Feasibility study into mechanisms for the support of encapsulated ISUP information in IMS
ETSI TS 183 043	IMS-based PSTN/ISDN Emulation Stage 3 specification PES Stage 3
ETSI TS 182 012	IMS-based PSTN/ISDN Emulation Subsystem; Functional architecture IMS-based Emulation
ETSI ES 283 031	IP Multimedia: H.248 Profile for controlling Multimedia Resource Function Processors (MRFP) in the IP Multimedia System (IMS); Protocol specification H.248 and MRFP
ETSI TS 183 033	IP Multimedia; Diameter based protocol for the interfaces between the Call Session Control Function and the User Profile Server Function/Subscription Locator Function; Signalling flows and protocol details
ETSI TS 183 021	Endorsement of 3GPP TS 29.162 Interworking between IM CN Sub-system and IP networks 29.162 endorsement
ETSI TR 182 005	Organization of user data
ETSI ES 283 030	Presence Service Capability; Protocol Specification
ETSI TS 182 008	Presence Service; Architecture and functional description Presence stage 2
ETSI TR 181 007	Overview of Messaging Services: Messaging Overview
ETSI TS 183 041	Messaging service using the IP Multimedia (IM) Core Network (CN) subsystem; Stage 3: Protocol specifications
ETSI ES 282 004	NGN Functional Architecture; Network Attachment Sub-System (NASS)

ETSI ES 283 035	Network Attachment Sub-System (NASS); e2 interface based on the DIAMETER protocol
ETSI ES 283 034	Network Attachment Sub-System (NASS); e4 interface based on the DIAMETER protocol
ETSI ES 282 003	Resource and Admission Control Sub-system (RACS); Functional Architecture
ETSI TS 183 017	Resource and Admission Control: DIAMETER protocol for session based policy set-up information exchange between the Application Function (AF) and the Service Policy Decision Function (SPDF); Protocol specification Gq' interface stage 3
ETSI ES 283 018	Resource and Admission Control: H.248 Profile for controlling Border Gateway Functions (BGF) in the Resource and Admission Control Subsystem (RACS); Protocol specification la Interface H.248
ETSI ES 283 026	Resource and Admission Control; Protocol for QoS reservation information exchange between the Service Policy Decision Function (SPDF) and the Access-Resource and Admission Control Function (A-RACF) in the Resource and Protocol specification Rq Interface
ETSI TS 185 001	Next Generation Network (NGN); Quality of Service (QoS) Framework and Requirements
ETSI TR 182 015	Next Generation Networks; Architecture for Control of Processing Overload NGN Control
ETSI ES 282 002	PSTN/ISDN Emulation Sub-system (PES); Functional architecture PES architecture
ETSI TS 183 004 – 0012,0015,0016	PSTN/ISDN Simulation Services : CDIV, CONF, MWI, OIP, TIP, CW, HOLD, ACR-CB, AoC, CCBS-CCNR, MCID: Protocol services
ETSI TS 183 023	PSTN/ISDN simulation services; Extensible Markup Language (XML) Configuration Access Protocol (XCAP) over the Ut interface for Manipulating NGN PSTN/ISDN Simulation Services
ETSI TS 181 010	Service requirements for end-to-end session control in multimedia networks (Release 1) E2E session
ETSI TS 181 005	Services and Capabilities Requirements
ETSI TR 181 003	Services Capabilities, Requirements and strategic direction for NGN services
ETSI TS 187 001	NGN SECurity (SEC); Requirements
ETSI TS 187 003	NGN Security; Security Architecture
ETSI TR 188 004	NGN Management; OSS vision

ETSI TS 188 001	NGN management; Operations Support Systems Architecture
ETSI TS 188 003	OSS requirements; OSS definition of requirements and priorities for further network management specifications for NGN
ETSI TR 102 647	Network Management; NGN Management standards; Overview and gap analysis
ETSI TS 182 028	IPTV Architecture; Dedicated subsystems for IPTV functions
ETSI TS 182 027	IPTV Architecture; IPTV functions supported by the IMS subsystem

PacketCable 2.0 Standards

The PacketCable 2.0 specifications can be obtained from the CableLabs website http://
www.cablelabs.com/specifications/pc20.html

PKT-SP-23.008-I02-070925	PacketCable™ 2.0 IMS Delta Specifications Organization of subscriber data Specification 3GPP TS 23.008
PKT-SP-24.229-I03-070925	PacketCable™ 2.0 IMS Delta Specifications Session Initiation Protocol (SIP) and Session Description Protocol (SDP); Stage 3 Specification 3GPP TS 24.229
PKT-SP-29.228-I02-070925	PacketCable™ 2.0 IMS Delta Specifications IP Multimedia (IM) Subsystem Cx and Dx Interfaces; Signalling Flows and Message Contents Specification 3GPP 29.228
PKT-SP-29.229-I02-070925	PacketCable™ 2.0 IMS Delta Specifications Cx/Dx Interfaces based on the Diameter Protocol Specification 3GPP TS 29.229
PKT-SP-33.203-I03-070925	PacketCable™ 2.0 IMS Delta Specifications 3G security; Access security for IP-based services Specification 3GPP TS 33.203
PKT-SP-33.210-I03-070925	PacketCable™ 2.0 IMS Delta Specifications 3G Security; Network Domain Security; IP network layer security Specification 3GPP TS 33.210
PKT-SP-ACCT-I03-070925	PacketCable™ Accounting Specification
PKT-SP-CODEC-MEDIA-I03-070925	PacketCable™ Codec and Media Specification
PKT-SP-CPD-I03-070925	PacketCable™ Control Point Discovery Interface Specification
PKT-SP-ES-DCI-I02-070925	PacketCable™ 2.0 PacketCable Electronic Surveillance Delivery Function to Collection Function Interface Specification

PKT-SP-ES-INF-I03-070925   PacketCable™ Electronic Surveillance Intra-Network Specification

PKT-SP-PRS-I01-070628   PacketCable™ 2.0 Presence Specification

PKT-SP-EUE-DATA-I01-071106   PacketCable™ 2.0 E-UE Provisioning Data Model Specification

PKT-SP-EUE-PROV-I01-071106   PacketCable™ 2.0 E-UE Provisioning Framework Specification

PKT-SP-QOS-I01-070925   PacketCable™ 2.0 Quality of Service Specification

OMA Standards

3GPP TS 22.127   Service Requirement for the Open Services Access (OSA); Stage 1

3GPP TS 23.198   Open Service Access (OSA); Stage 2

3GPP TS 23.198-[01-16]   OSA APIs

3GPP TS 23.199-[01-20]   OSA; Parlay X web services

ITU Standards

ITU Rec. E.164   The international public telecommunication numbering plan.

ITU Rec. G.711   Pulse code modulation (PCM) of voice frequencies.

ITU Rec. H.248   Gateway control protocol.

ITU Rec. H.263   Annex X (03/04): Annex X: Profiles and levels definition.

ITU Rec. H.264   Advanced video coding for generic audiovisual services.

ITU Rec. T.120   Data protocols for multimedia conferencing.

ITU Rec. T.126   Multipoint still image and annotation protocol.

ITU Rec. T.128   Multipoint application sharing.

IETF RFCs

IMS standards also refer to Internet Engineering Task Force (IETF) Request for Comments (RFC). These standards can be accessed from the ietf website at http://www. ietf.org/rfc/rfc*nnnn*.txt, where *nnnn* is the number of the RFC. The drafts are work in progress.

RFC0791   Internet Protocol, STD 5. (September 1981)

RFC0793   Transmission Control Protocol, STD 57. (September 1981)

RFC2131   Dynamic host configuration protocol. (March 1997)

RFC2132   DHCP Options and BOOTP Vendor Extensions. (March 1997)

RFC2401   Security Architecture for the Internet Protocol. (November 1998)

RFC2462   IPv6 Address Autoconfiguration. (November 1998)

RFC2486   The Network Address Identifier. (January 1999)

RFC2617   HTTP Authentication: Basic and Digest Access Authentication. (June 1999)

RFC2833   RTP Payload for DTMF Digits, Telephony Tones and Telephony Signals. (May 2000)

RFC2960   Stream Control Transmission Protocol. (October 2000)

RFC2976   The SIP INFO method. (October 2000)

RFC3041   Privacy Extensions for Stateless Address Autoconfiguration in IPv6. (January 2001)

RFC3261   SIP: Session Initiation Protocol. (June 2002)

RFC3262   Reliability of provisional responses in Session Initiation Protocol (SIP). (June 2002)

RFC3263   Session Initiation Protocol (SIP): Locating SIP Servers. (June 2002)

RFC3264   An Offer/Answer Model with Session Description Protocol (SDP). (June 2002)

RFC3265   Session Initiation Protocol (SIP) Specific Event Notification. (June 2002)

RFC3310   Hypertext Transfer Protocol (HTTP) Digest Authentication Using Authentication and Key Agreement (AKA). (September 2002)

RFC3311   The Session Initiation Protocol (SIP) UPDATE method. (September 2002)

RFC3312   Integration of resource management and Session Initiation Protocol (SIP). (October 2002)

RFC3313   Private Session Initiation Protocol (SIP) Extensions for Media Authorization. (January 2003)

RFC3315   Dynamic Host Configuration Protocol for IPv6 (DHCPv6). (July 2003)

RFC3319   Dynamic Host Configuration Protocol (DHCPv6) Options for Session Initiation Protocol (SIP) Servers. (July 2003)

RFC3320   Signaling Compression (SigComp). (March 2002)

RFC3323   A Privacy Mechanism for the Session Initiation Protocol (SIP). (November 2002)

RFC3325   Private Extensions to the Session Initiation Protocol (SIP) for Network Asserted Identity within Trusted Networks. (November 2002)

RFC3326   The Reason Header Field for the Session Initiation Protocol (SIP). (December 2002)

RFC3327   Session Initiation Protocol Extension Header Field for Registering Non-Adjacent Contacts. (December 2002)

RFC3329   Security Mechanism Agreement for the Session Initiation Protocol (SIP). (January 2003)

RFC3361   Dynamic Host Configuration Protocol (DHCP-for-IPv4) Option for Session Initiation Protocol (SIP) Servers. (August 2002)

RFC3388   Grouping of Media Lines in Session Description Protocol. (December 2002)

RFC3420   Internet Media Type message/sipfrag. (November 2002)

RFC3428   Session Initiation Protocol (SIP) Extension for Instant Messaging. (December 2002)

RFC3455   Private Header (P-Header) Extensions to the Session Initiation Protocol (SIP) for the 3rd-Generation Partnership Project (3GPP). (January 2003)

RFC3485   The Session Initiation Protocol (SIP) and Session Description Protocol (SDP) static dictionary for Signaling Compression (SigComp). (February 2003)

RFC3486   Compressing the Session Initiation Protocol (SIP). (February 2003)

RFC3515   The Session Initiation Protocol (SIP) REFER method. (April 2003)

RFC3524   Mapping of Media Streams to Resource Reservation Flows. (April 2003)

RFC3550   RTP: A Transport Protocol for Real-Time Applications. (July 2003)

RFC3550   RTP: A Transport Protocol for Real-Time Applications. (July 2003)

RFC3556   Session Description Protocol (SDP) Bandwidth Modifiers for RTP Control Protocol (RTCP) Bandwidth. (July 2003)

RFC3581   An Extension to the Session Initiation Protocol (SIP) for Symmetric Response Routing. (August 2003)

RFC3588   Diameter Base Protocol. (September 2003)

RFC3589   Diameter Command Codes for 3GPP. (September 2003)

RFC3603   Private Session Initiation Protocol (SIP) Proxy-to-Proxy Extensions for Supporting the PacketCable Distributed Call Signaling Architecture. (October 2003)

RFC3608   Session Initiation Protocol (SIP) Extension Header Field for Service Route Discovery During Registration. (October 2003)

RFC3646   DNS Configuration options for Dynamic Host Configuration Protocol for IPv6 (DHCPv6). (December 2003)

RFC3680   A Session Initiation Protocol (SIP) Event Package for Registrations. (March 2004)

RFC3761   The E.164 to Uniform Resource Identifiers (URI) Dynamic Delegation Discovery System (DDDS) Application (ENUM). (April 2004)

RFC3825   Dynamic Host Configuration Protocol Option for Coordinate-based Location Configuration Information. (July 2004)

RFC3840   Indicating User Agent Capabilities in the Session Initiation Protocol (SIP). (August 2004)

RFC3841   Caller Preferences for the Session Initiation Protocol (SIP). (August 2004)

RFC3842   A Message Summary and Message Waiting Indication Event Package for the Session Initiation Protocol (SIP). (August 2004)

RFC3856   A Presence Event Package for the Session Initiation Protocol (SIP). (August 2004)

RFC3857   A Watcher Information Event Template Package for the Session Initiation Protocol (SIP). (August 2004)

RFC3861   Address Resolution for Instant Messaging and Presence. (August 2004)

RFC3891   The Session Initiation Protocol (SIP) Replaces Header. (September 2004)

RFC3892   The Session Initiation Protocol (SIP) Referred-By Mechanism. (September 2004)

RFC3903   An Event State Publication Extension to the Session Initiation Protocol (SIP). (October 2004)

RFC3911   The Session Initiation Protocol (SIP) Join Header. (October 2004)

RFC3948   UDP Encapsulation of IPsec ESP Packets. (January 2005)

RFC3966   The tel URI for Telephone Numbers. (December 2004)

RFC3984   RTP Payload Format for H.264 Video. (February 2005)

RFC3986   Uniform Resource Identifiers (URI): Generic Syntax. (January 2005)

RFC4028   Session Timers in the Session Initiation Protocol (SIP). (April 2005)

RFC4032   Update to the Session Initiation Protocol (SIP) Preconditions Framework. (March 2005)

RFC4077   A Negative Acknowledgement Mechanism for Signaling Compression. (May 2005)

RFC4119   A Presence-based GEOPRIV Location Object Format. (December 2005)

RFC4145   TCP-Based Media Transport in the Session Description Protocol (SDP). (September 2005)

RFC4168   The Stream Control Transmission Protocol (SCTP) as a Transport for the Session Initiation Protocol (SIP). (October 2005)

RFC4244   An Extension to the Session Initiation Protocol (SIP) for Request History Information. (November 2005)

RFC4320   Actions Addressing Identified Issues with the Session Initiation Protocol's (SIP) Non-INVITE Transaction. (January 2006)

RFC4346   The TLS Protocol Version 1.1. (April 2006)

RFC4354   A Session Initiation Protocol (SIP) Event Package and Data Format for Various Settings in Support for the Push-to-Talk over Cellular (PoC) Service. (January 2006)

RFC4411   Extending the Session Initiation Protocol (SIP) Reason Header for Preemption Events. (February 2006)

RFC4412   Communications Resource Priority for the Session Initiation Protocol (SIP). (February 2006)

RFC4457   The Session Initiation Protocol (SIP) P-User-Database Private-Header (P-header). (April 2006)

RFC4458   Session Initiation Protocol (SIP) URIs for Applications such as Voicemail and Interactive Voice Response (IVR). (January 2006)

RFC4566   SDP: Session Description Protocol. (June 2006)

RFC4575   A Session Initiation Protocol (SIP) Event Package for Conference State. (August 2006)

RFC4583   Session Description Protocol (SDP) Format for Binary Floor Control Protocol (BFCP) Streams. (November 2006)

RFC4662   A Session Initiation Protocol (SIP) Event Notification Extension for Resource Lists. (August 2006)

RFC4694   Number Portability Parameters for the 'tel' URI. (October 2006)

RFC4694   The P-Answer-State Header Extension to the Session Initiation Protocol for the Open Mobile Alliance Push to Talk over Cellular. (September 2007)

RFC4769   IANA Registration for an Enum service Containing Public Switched Telephone Network (PSTN) Signaling Information. (November 2006)

RFC4825   The Extensible Markup Language (XML) Configuration Access Protocol (XCAP). (May 2007)

RFC4867   RTP payload format and file storage format for the Adaptive Multi-Rate (AMR) Adaptive Multi-Rate Wideband (AMR-WB) audio codecs. (April 2007)

RFC4867   RTP Payload Format and File Storage Format for the Adaptive Multi-Rate (AMR) and Adaptive Multi-Rate Wideband (AMR-WB) Audio Codecs. (April 2007)

RFC4896   Signaling Compression (SigComp) Corrections and Clarifications Implementer's Guide for SigComp. (June 2007)

RFC4967   Dial String Parameter for the Session Initiation Protocol Uniform Resource Identifier. (July 2007)

RFC4975   The Message Session Relay Protocol. (September 2007)

RFC5002   The Session Initiation Protocol (SIP) P-Profile-Key Private Header (P-Header). (August 2007)

RFC5009   Private Header (P-Header) Extension to the Session Initiation Protocol (SIP) for Authorization of Early Media. (September 2007)

draft-drage-sipping-service-identification-01	A Session Initiation Protocol (SIP) Extension for the Identification of Services. (July 2007)
draft-garcia-simple-presence-dictionary-06	The Presence-Specific Static Dictionary for Signaling Compression (Sigcomp). (August 2007)
draft-ietf-behave-rfc3489bis-09	Session Traversal Utilities for (NAT) (STUN). (August 2007)
draft-ietf-behave-turn-04	Traversal Using Relays around NAT (TURN): Relay Extensions to Session Traversal Utilities for NAT (STUN). (July 2007)
draft-ietf-ecrit-service-urn-06	A Uniform Resource Name (URN) for Services. (March 2007)
draft-ietf-rohc-sigcomp-sip-07	Applying Signaling Compression (SigComp) to the Session Initiation Protocol (SIP). (July 2007)
draft-ietf-sip-acr-code-05	Rejecting Anonymous Requests in the Session Initiation Protocol (SIP). (July 2007)
draft-ietf-sip-consent-framework-02	A Framework for Consent-Based Communications in the Session Initiation Protocol (SIP). (July 2007)
draft-ietf-sip-fork-loop-fix-05	Addressing an Amplification Vulnerability in Session Initiation Protocol (SIP) Forking Proxies. (March 2007)
draft-ietf-sip-gruu-14	Obtaining and Using Globally Routable User Agent (UA) URIs (GRUU) in the Session Initiation Protocol (SIP). (June 2007)
draft-ietf-sip-location-conveyance-08	Session Initiation Protocol Location Conveyance. (July 2007)
draft-ietf-sip-multiple-refer-01	Referring to Multiple Resources in the Session Initiation Protocol (SIP). (July 2007)
draft-ietf-sip-outbound-10	Managing Client Initiated Connections in the Session Initiation Protocol (SIP).
draft-ietf-sipping-config-framework-12	A Framework for Session Initiation Protocol User Agent Profile Delivery. (May 2007)
draft-ietf-sipping-gruu-reg-event-09	Reg Event Package Extension for GRUUs. (July 2007)
draft-ietf-sip-uri-list-conferencing-01	Conference Establishment Using Request-Contained Lists in the Session Initiation Protocol (SIP).
draft-ietf-sip-uri-list-message-01	Multiple-Recipient MESSAGE Requests in the Session Initiation Protocol (SIP).

draft-ietf-sip-uri-list-subscribe-01   Subscriptions to Request-Contained Resource Lists in the Session Initiation Protocol (SIP).

draft-jesske-sipping-etsi-ngn-reason-02   Use of the Reason header field in Session Initiation Protocol (SIP) responses. (January: 2008)

Research Papers

[R1]   Abhaya Asthana et al., End-to-End Availability Considerations for Services Over IMS Bell Labs Technical Journal 11(3), 199–210 (2006)

[R2]   Adiseshu Hari, Intelligent Media gateway Selection in Carrier-Grade VoIP Networks, Bell Labs Technical Journal Bell Labs Technical Journal 10(4), 133–150 (2006)

[R3]   Ake Gustafsson, Enriched messaging architecture, Ericsson Review No. 2, 2007

[R4]   Amjad Akkawi et al., A Mobile Gaming Platform for the IMS, Proceedings of 3rd ACM SIGCOMM workshop on Network and system support for games, 2004

[R5]   Andre Beck, et al., Blending Telephony and IPTV, Bell Labs Technical Journal 12(1), 23–39 (2007)

[R6]   Andy Johnston et al., Evolution of service delivery platforms, Ericsson Review No. 1, 2007

[R7]   Carlos Urrutia-Valdés, Presence and Availability with IMS: Applications Architecture, Traffic Analysis, and Capacity Impacts, Bell Labs Technical Journal 10(4), 101–107 (2006)

[R8]   Christina Birkehammar, New high-quality voice service for mobile networks, Ericsson Review No. 3, 2006

[R9]   Daniel Enström et al., Multimedia telephony for IMS – Interoperable VoIP with multimedia support, Ericsson 44 Review No. 2, 2007

[R10]   Daniel F. Lieuwen et al., Subscriber Data Management in IMS Networks, Bell Labs Technical Journal 10(4), 197–215 (2006)

[R11]   Denise M. Ward et al., Domain Management of IMS, Bell Labs Technical Journal 10(4), 233–254 (2006)

[R12]   Dupyo Choi et al., Transition to IPv6 and Support for IPv4/IPv6 Interoperability in IMS, Bell Labs Technical Journal 10(4), 261–270 (2006)

[R13]   Erik E. Anderlind et al., IMS Security, Bell Labs Technical Journal 11(1), 37–58 (2006)

[R14]   Frank M. Alfano, IMS Service-Based Bearer Control, Bell Labs Technical Journal 10(4), 151–166 (2006)

[R15]   IMS: Application Enabler and UMTS/HSPA Growth Catalyst, 3G Americas, July 2006

[R16]   Jacco Brok, Enabling New Services by Exploiting Presence and Context Information in IMS, Bell Labs Technical Journal 10(4), 83–100 (2006)

[R17]   Joachim Fabini et al., "IMS in a Bottle": Initial Experiences from an OpenSER-based Prototype Implementation of the 3GPP IP Multimedia Subsystem, Proc. International Conference on Mobile Business (ICMB), June 2006.

[R18]   John Heck, Blended IMS Messaging Applications, Bell Labs Technical Journal 10(4), 39–52 (2006)

[R19]   Justino Santos et al., Multicast/Broadcast Network Convergence in Next Generation Mobile Networks, Computer Networks: The International Journal of Computer and Telecommunications Networking, Jan 2008

[R20]   Kristin F. Kocan et al., A Novel Software Approach for Service Brokering in Advanced Service Architectures, Bell Labs Technical Journal 11(1), 5–20 (2006)

[R21]   Luay H. Tahat et al., Dual Mode Service: Learning From Integrating an IP Multimedia Subsystem Service Into Existing Live Networks, Bell Labs Technical Journal 11(4), 135–150 (2007)

[R22]   Maria R. G. Azada et al., Seamless Mobility Across IMS and Legacy Circuit Networks, Bell Labs Technical Journal 10(4), 25–38 (2006)

[R23]   Mehdi Mani, Access to IP multimedia subsystem of UMTS via PacketCable network, Wireless 2005 IEEE Communications and Networking Conference, March 2005

[R24]   Michael R. Brenner, Service-Oriented Architecture and Web Services Penetration in Next-Generation Networks, Bell Labs Technical Journal 12(2), 147–160 (2007)

[R25]   Miran Mosmondor et al., Conveying and handling location information in the IP Multimedia Subsystem

[R26]   Nicholas M. DeVito et al., Functionality and Structure of the Service Broker in Advanced Service Architectures, Bell Labs Technical Journal 10(1), 17–30 (2005)

[R27]   Nick J. Mazzarella, Advantages of Harmonized IMS-Based Charging Architecture in Different AccessTechnologies, Bell Labs Technical Journal 10(4), 109–115 (2006)

[R28]   Piotr Kessler, Ericsson IMS Client Platform, Ericsson Review No. 2 2007

[R29]   Ramesh Pattabhiraman, Enhanced Active Phone Book Services: Blended Lifestyle Services Made Real!, Bell Labs Technical Journal 11(4), 315–326 (2007)

[R30]   Robert Gaglianello, IMS Shared Streaming Video, Bell Labs Technical Journal 10(4), 71–75 (2006)

[R31]   Sheng Chen, IP Multimedia Subsystem Converged Call Control Services, Bell Labs Technical Journal 12(1), 145–160 (2007)

[R32]   Technical Specification of IMS weShare, Ericsson, 20/287 01-FGB 101 95 Uen Rev A

[R33]   The Evolution of UMTS/HSDPA - 3GPP Release 6 and Beyond, 3G Americas, December 2005

[R34]   Thierry Bessis, Improving Performance and Reliability of an IMS Network by Co-Locating IMS Servers, Bell Labs Technical Journal 10(4), 167–178 (2006)

[R35]   Thomas Magedanz et al., Evolution of SOA Concepts in Telecommunications, IEEE Computer, November 2007

[R36]   Torsten Dinsing et al., Service composition in IMS using Java EE SIP servlet containers, Ericsson Review No.3 2007

[R37]   Tsunehiko Chiba et al., Gap Analysis and Deployment Architectures for 3GPP2 MMD Networks, IEEE Vehicular Technology Magazine, March 2007

[R38]   Wenhua Jiao, Provisioning End-to-End QoS Under IMS Over a WiMAX Architecture, Bell Labs Technical Journal 12(1), 115–121 (2007)

[R39]   William J. Barnett, Jr., Enabling New Service Provider Business Models with the IP Multimedia Subsystem, Bell Labs Technical Journal 10(4), 7–15 (2006)

[R40]   Yigang Cai et al., IP Multimedia Subsystem Online Session Charging Call Control, Bell Labs Technical Journal 10(4), 117–132 (2006)

[R41]   Ying Hu et al., IMS Service Enhancement Layer: A Quantitative Value Proposition, Bell Labs Technical Journal 12(1), 95–114 (2007)

Books

[R41]   Gonzalo Camarillo, Miguel A. Garcia Martin, The 3G IP Multimedia Subsystem, Second Edition, John Wiley

[R42]   Mikka Poikselka et al., The IMS IP Multimedia Concepts and Services, Second Edition, John Wiley

[R43]   Chad Hart, Ensuring Quality in IMS: A Guide to Testing and Monitoring IMS Products, Networks, and Services, Empirix

[R44]   Colin Perkins, RTP: Audio and Video for the Internet, Addison Wesely

[R45]   Dave Wisely et al., IP for 3G Networking Technologies for Mobile Communications, John Wiley

[R46]   Heikki Kaaranen et al., UMTS Networks: Architecture, Mobility and Services, John Wiley

[R47]   Henry Sinnreich (Author), Alan B. Johnston, Internet Communications Using SIP: Delivering VoIP and Multimedia Services with Session Initiation Protocol, John Wiley

[R48]   Juha Korhonen, Introduction to 3G Mobile Communications, Second Edition, John Wiley

[R49]   Madjid and Mahsa Nakhjiri, AAA and Network Security for Mobile Access, John Wiley

Web Resources

- http://www.tech-invite.com/ is a valuable resource for standards navigation.

- The http://www.ims-lantern.com blog provides insight into emerging areas of IMS

- The http://opengardensblog.futuretext.com blog discusses user aspects of IMS

- The http://www.imsquality.com blog focuses on quality and testing of IMS

- Market Research Data about IMS
  - ABI Research: http://www.abiresearch.com
  - Unstrung: http://www.unstrung.com
  - LightReading: http://www.lightreading.com
  - VDC: http://www.vdc-corp.com
  - 3GPP Americas: http://www.3gamericas.com

- The online journal http://telephonyonline.com/ims/provides news update on IMS

- Software Downloads for IMS

  - Open Source software code from the Fraunhofer FOKUS Labs http://www.openimscore.org/
  - Basic HSS implementation—http://code.google.com/p/hss/

Software development kit for proof of concept IMS applications and testing http://www.ericsson.com/mobilityworld/sub/open/technologies/ims_poc/tools/sds_40

Index

Printed in the United States
By Bookmasters